Ebadollah Amouzad Mahdiraji

Hybrid Renewable Energy System Optimization

Ebadollah Amouzad Mahdiraji

Hybrid Renewable Energy System Optimization

Optimization of Hybrid Renewable Energy

Noor Publishing

Publisher:
Noor Publishing
is a trademark of
International Book Market Service Ltd., member of OmniScriptum Publishing Group
17 Meldrum Street, Beau Bassin 71504, Mauritius
Printed at: see last page
ISBN: 978-620-0-77535-1

Hybrid Renewable Energy System Optimization

Ebadollah Amouzad Mahdiraji

Department of Engineering, Sari Branch, Islamic Azad University, Sari, Iran

Email: ebad.amouzad@gmail.com

This Book is dedicated to

My Father and Mother

Those who are the best in our life

Table of Content

Chapter I

Introduction

Chapter II

Introducing Components in the Structure of a Hybrid Network

Chapter III

Simulation Method and Optimal Measurement Method

Chapter IV

Simulation Results and Analysis

Chapter V

Conclusions

References

Figures

Tables

Chapter I

Introduction

Introduction

Optimizing a hybrid system that includes photovoltaic panels, a backup source (micro-turbine or diesel) and a battery system minimizes the cost of power generation. In this book, we present an outline of this optimal system that supplies electricity or power. A scenario that depends on a standalone PV and the other a backup source. This optimization, obtained through the use and application of a genetic algorithm, minimizes the COE consumption function while covering the load demand with the specified value for the probability of loss. The costs of publishing a global ad are included in this optimization analysis. First, the solar radiation data is analyzed and optimized, and the angle of inclination of the PV panels is optimized. Powering a small area using this hybrid system is cost-effective and very useful compared to expanding the facility network to this remote area or just using conventional resources for this purpose. This hybrid system reduces operating costs and emissions. A hybrid system that recognizes these optimization goals is a system made of combining these resources. Hybrid energy systems that rely on renewable energies, especially solar photo-voltaics, are widely used today. Their effectiveness has been proven when used to store power in various locations, especially for small isolated loads. Their use can mitigate the effects of greenhouse gases such as Co_2, No, No_2 and So_2 to comply with Kyoto. The low cost of maintaining them and releasing low pollutants are the best advantages. [8-1]

This book considers a method based on a genetic algorithm to design a solar PV hybrid system as a renewable source and a small turbine or diesel generator as its backup source. The use of systems with more than one energy source is known as hybrid systems in energy sources that can increase the reliability and security of energy compared to systems with only one energy source. [12-9]

Today, the world has shifted to sustainable and clean energy sources due to the reduction of fossil fuels such as oil and gas and environmental issues due to the use of these energy sources such as increased global warming, increased pollutants and greenhouse gases such as carbon dioxide. Types of renewable energy are wind, solar, hydroelectric, biomass and thermal energy. The combination of renewable energy sources with conventional sources has been used to form a hybrid system to achieve desirable high-power energy sources and low costs and reduce transmission losses, etc. [8, 9] At an economical price and based on the reliability of the hybrid system, its design must be optimized in terms of operation and selection of key components, so an optimal measurement method is very necessary to be useful and economical for the hybrid system. Optimal measurement of these systems requires careful analysis of a specific parameter due to the influence of various situations such as solar radiation, wind speed, and temperature and their relationship to the time-dependent cost of the system. Various optimization techniques such as genetic algorithm, particle batch optimization, etc. are used for this purpose. In this book, a genetic algorithm will be used to optimize different component sizes of a small photovoltaic-turbine hybrid system in a sample micro grid. Here, optimization components of various types are used to minimize the objective function of energy cost. Optimization parameters include photovoltaic deflection angles and surface orientation as well as the type of installation.

1.1. Purpose of research

The main objective of this research is to investigate the existing structures for optimizing the components of a hybrid system and thus minimizing the cost of energy production.

Types of hybrid systems depend more on the availability and availability of renewable energy. In the study area, solar radiation has a high potential for annual sunlight hours. Mean values between 5.5 Kw/hm^2 and 6 Kw/hm^2 were recorded at a horizontal level for daily mean daily solar radiation. [13]

There are many small communities in remote areas that rely on diesel generators as their home's electricity source. It is also dependent on the external energy source and has limitations on the capacity required. As a result, more than 30 percent of households receive a cut off energy source [14]. Providing these remote locations to solve energy shortage issues is the most important priority of various institutions. Small projects have been implemented using diesel/PV hybrid systems to power energy in parts of these small communities. Small turbines have several advantages over diesel generators: lower operating and maintenance costs, higher levels of reliability, lower noise and less pollutant emissions and greater fuel flexibility [9]. These features increase the use of small turbines as standby sources compared to diesel generators.

Balanced energy for each hour throughout the year is a step that must be considered before performing or performing the optimization. For this purpose, the energy generated by each energy source must be calculated. This requires the mathematical modeling of any component or component in the system and the availability of atmospheric data, which requires the analysis of this data.

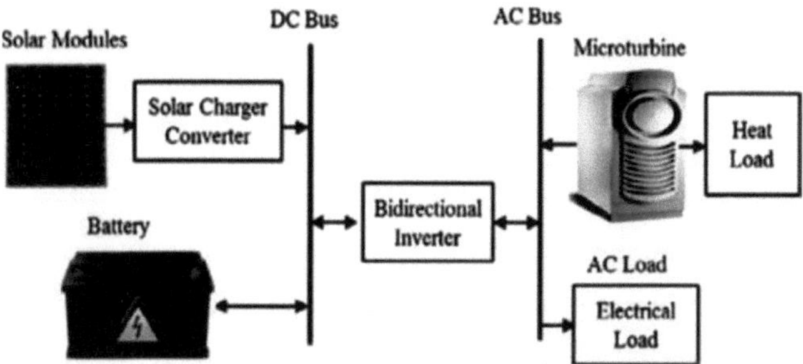

Figure -1-1 Block diagram of the proposed hybrid system

Solar radiation analysis and the development of a mathematical model of the components that constitute the hybrid system are essential steps that must be completed before optimizing and completing the components of the hybrid system that will be presented at the end of the project. [13, 15] To analyze solar radiation. Calculate the angles of the surface and the tilt, predict the solar radiation, and predict the solar panels using the mathematical model. These developed methods are used in this research. While duplicate or graphical techniques have been recommended and used to optimize hybrid system components by many researchers, a new algorithm has been used by a smaller number of researchers for this purpose. In new ways, a design space is constructed from possible solutions for optimal component sizes from the beginning. A search method is also used to select the most optimal formatting to match the target functions. These methods are recommended when it is appropriate for the multi-objective function or when different parameters (variables) make a decision vector that needs to be optimized. They are also recommended when the hybrid system operating strategy needs to be optimized. In fact, using linear descriptive programs or duplicate methods to solve these optimization problems is difficult. Genetic algorithms, particle batch optimization, and simulation plating are examples of these new search techniques. Less common methods may be found in history, and each of these methods has its own characteristics, but a review of history shows that the genetic algorithm has excellent properties that make it ideal for optimization purposes, especially for hybrid systems. Its performance "is very good for general search optimization and is suitable for many optimization parameters optimization problems. Genetic algorithms have relatively more difficult coding than other algorithms and require more computation time to solve optimization problems [16].

In this book, the genetic algorithm was used to optimize the sizes of different components of the hybrid system, where the variants of these components were selected from different types and minimized the COE objective function. Optimized parameters also include PV tilting angles and surface area. Here's a chance to optimize the type of each component. In this case, the type of PV installation structure is also optimized. Comparison of the results of this hybrid system with the hybrid system in which a diesel generator is used as a backup source instead of a small turbine. The effect on COE has been studied concerning the case that the load is covered with 100% reliability and that a certain value is determined for the LLP.

Chapter II

Introducing Components in the Structure of a Hybrid Network

2.1. Diesel Generator

Diesel generators are a combination of diesel engines, generators and a variety of accessories such as chassis, overhead compartment, sound insulation, control devices, emergency circuit breakers, heat-generating system, automatic starter systems, etc. used to generate electricity.

2.1.1. Diesel Generator machine

Engine generators can generate from 1 to 4KVA for homes, shops, small offices and up to (1KVA / 1MVA) usable for large office complexes and factories. A 1KV generator can be housed in a removable isolation chamber. 2MW generators are used for small power stations and several generators can be used. The larger generators are shipped separately to the installation site, where assembly and accessories are added.

Small diesel generators up to 4KV are not only used to generate uninterrupted power, but also to supply the electricity needed continuously either at peak power consumption or even when there is a shortage of larger generators.

Ships and many large land vehicles, such as trains, also use diesel generators not only to provide light but also to provide the driving force they need. Electric motors can move more smoothly and make more efficient use of space. Electric motors were used in ships before World War I and evolved during World War II.

Electricity generating units are classified and named in terms of generating capacity, maximum output, and in KW, depending on the type of consumption used for continuous or emergency power generation.

It is a combination of a diesel engine and an electric generator (often an alternator) used to convert electrical energy. Diesel generator sets are used in places without a power grid connection, such as when there is a pressing need for power and when there is no grid. When the network is disconnected, more sophisticated applications such as peak shear, network support and transmission to power networks are provided.

Considering the size of the diesel generator is essential to prevent overload or power failure and this has been made difficult by modern electronics, especially nonlinear loads. The electric charge generator set includes all the features of KW, KVA, VAR and harmonic contents including the starting currents (generally from the motor) and the nonlinear loads. The expected task, for example, emergency, continuous or initial power (prime), should be well considered in environmental conditions such as altitude, temperature, and greenhouse gas regulation. Many manufacturers of larger generators will provide a suite of software that will perform complex size calculations by simply reading site conditions and connected electrical charge characteristics.

2.1.2. Some of the Most Common Motors

CUMMINS/ VOLVO/ GM/ PERKINS/ DOOSAN/ MTU/ Mitsubishi

2.1.3. Some of the Most Common Generators

STAMFORD / STAMFORD POWER / MECC ALTE

One or more diesel generators operating without connection to the power grid operator in the so-called island state, which provides the features of several parallel generators that have the advantages of redundancy and better efficiency at partial loads. This plant puts the generator online and switching off depends on the system and time.

An island power plant intended for the main power source will have a separate community of at least three diesel generators, both of which are expected to carry the load required. Groups are not uncommon up to 8. Generators can be electrically connected through simultaneous processing. This synchronization involves adjusting the voltage, frequency, and phase before the generator is connected to the system. Loss of synchrony before the connection can cause a short-circuit upstream or wear and tear on the generator or panel. The synchronization process can be done automatically by an automated synchronizer module. The load can be shared between the generators running in parallel through load sharing. The load splitting can be controlled by the frequency drop control by the frequency in the generator, while the motor fuel control is continuously adjusted for load change and remains from power sources.

A diesel generator gets more load when the fuel supply for combustion is increased and the load will be released if the fuel supply is reduced.

In addition to well-known tasks such as powering off the grid, diesel generators typically support the main power grids worldwide in two separate ways.

2.1.4. Network Support

Emergency-ready diesel generators and hydroelectric power plants, as a secondary function, are widely used to protect different national grids at any one time and for various reasons in the US and UK. This is always the case when, for example, a large capacity of about 4MW of power is suddenly lost or a sudden unintended jump in power causes spontaneous stored erosion. This is useful for both sections. Types of diesel have been purchased for other reasons as well, but they need to be thoroughly tested to make sure. The parallel network is the easiest way to do this. This method of operation is generally performed by a third party who manages the generator's performance and communicates with the system operator.

In this way, the British National Grid will operate at 5GW of power from a higher power plant and operate at the same rate for two minutes in some cases in parallel. This is much faster than the base power load, which can take up to 4 hours to cool down, and is faster than a gas turbine that can take a lot of time.

The fuel used for diesel is the fuel tested in the experiments. A similar system operating in France, with the maximum grid tariffs, will be equipped with at least 1GW of diesel generator sets to make it available. The primary job of diesels is to power the grid. During normal operation synchronized with the power plant network, they are operated with a 5% control speed drop. This means that all load speeds are 100% and no freight speeds are 105% and this requires stable network performance without the search and lack of power plants. Speed changes are generally the slowest. The power output settings are created by increasing the curve slope by increasing the spring pressure on a centrifugal governor. Generally, this is the basic system requirement for all power plants because new and old power plants must be adaptable in response to instantaneous changes in frequency without external communication dependence. The primary priority for generator support is to be taken into account. Specific uses are different, but a new diesel plant will consume between 0.4 and 0.28 liters of fuel per kilowatt-hour at generator terminals.

However, diesel engines can operate on a wide variety of fuels and depend on configuration through diesel fuel, which is generally obtained from crude oil. Engines can work with all spectrum of integrated fuel extracts from natural gas, alcohol, diesel, wood gas to diesel fuel from diesel fuel to sedimentary fuels. As such, they are introduced as consumable air and with a small amount of fossil fuel for combustion. Converting to 2% of fossil fuels can take place immediately.

2.1.5. Types of generators

By application, generator manufacturers offer four types:

- Standby
- Prime
- Continuous
- Multipurpose

The generator must provide the expected power required without any damage to the generator by giving one or more ratings.

Standby generator: Automatically resupply subscribers in a power outage. A model generator standby generator may need to operate for only a few hours a month, but other models of prime generators need to be operated continuously. When an alternator generator is running, it may operate under special circumstances, such as an additional 10 percent of the load that can occur during operation. A generator with the same model can launch a higher-order alternative service, which is for continuous operations. Manufacturers rank each generator according to the international agreement.

These standard ratings are defined to allow for proper selection and good comparisons between manufacturers to prevent machine malfunction and to guide designers.

Prime generator: They usually work in places where temporary electricity is needed. Like camps, exhibitions, mining, and camps. Should not be used in power applications. The output with different loads is unlimited for a single time. Generally, peak demand demands 100 percent of the initial ekW rating with 10 percent of extra load capacity for use in essential situations for a maximum of 1 hour to 12 hours. Ten percent of overload capacity is available for a limited time. Applications, where the generator is the sole source of power, are telecommunications to mines, construction sites, exhibition grounds and festivals, and so on.

Continuous generator: For power plants or electricity, issues are all over the world. It can be used to provide continuous power for a fixed time to full output rank for unlimited times. No substitution overload the generator is not available for this place at the same time to set up continuous or continuous parallel power and for the maximum permissible level, 8760 hours per year are located. It is also possible to cut a peak/support network even though the applied for 200 hours per year occur.

Multipurpose generator: It is recommended not to use this type of generator for economical applications. For example, there is no need for a permanent generator to have an automatic control panel at the time of a power outage. However, it is usually a common assumption that if a generator is 1000 KW in emergency mode, it will reach 850 and 800kW in the permanent state. Generally, how to load a generator is included in the generator specifications.

2.1.6. Motor failures

Diesel generators can cause adverse effects such as internal deposition (often leading to deposition in holes and columns) and carbon accumulation.

Ideally, diesel engines should run at least 60% to 75% times their maximum load. Short periods are running low times, allowing the collection to be fully loaded or near full load regularly.

Internal sediments and carbon accumulation are due to long periods of run at low speeds or low loads. Such a situation may occur if an engine releases its idle unit as a "standby" converter unit and is ready to arrive quickly when it is needed if the power to the set is applied to the load for which it is applied. This causes the diesel unit to be under load or, in many cases, when the set is started to operate, the load is discharged as an experiment. Starting a low-load engine causes low cylindrical pressures and as a result, the piston rings are weakly closed because it depends on the pressure of the gas against the oil film on the holes for forming the seals. Low cylinder pressure causes poor combustion, low combustion pressure, and temperature.

Incomplete combustion results in the formation of soot and residual unburned fuel, which closes and adheres to the piston rings, drastically reducing sealing efficiency and intensifying low elemental pressure. Sediments are created when the hot combustion gases burn poorly and block the piston ring, requiring lubrication on the cylindrical walls due to "sudden fuel", creating glaze-like fumes that drain the hole.

Hard carbon is formed from poor combustion and is very abrasive and abrasive, then increases fuel consumption (blue smoke), and is expected to maintain piston blockage and pressure.

Then the unburned fuel leaks through the rear of the piston ring and lubricates. Incomplete combustion causes the injectors to be blocked by soot and further aggravates combustion and black smoke. The problem caused by the formation of acids in engine oil is caused by concentrated water and combustion by-products that must be boiled normally at higher temperatures. This acidic build-up in lubricants causes slow but extremely destructive wear on the bearing surface.

The downturn cycle means that the engine will soon be irreversibly damaged and may never start and will not be able to achieve all the power it needs when it lasts longer.

Running under loads will inevitably not only cause white smoke from unburned fuel but over time will result in leaky blue smoke leaking from damaged old piston rings and black smoke resulting from injector failure. This contamination is not suitable for and adjacent. The creation of sediments or carbon has occurred. This can be remedied by removing the engine and re-penetrating the cylinder holes, cancellations, cleaning combustion chambers, fuel injection nozzles, and valves. If detected early, the engine starts at its maximum load to increase the internal pressure and temperature allowing the piston rings to hardly precipitate and allows the gradual accumulation of carbon to burn.

However, if these sediments progress, it is up to the stage to capture the piston rings inside their grooves. This will not affect. This can be prevented by carefully selecting the generator set by the manufacturer's printed instructions.

For a set of generators that have emergency use, the use of a supported load for testing may be impractical. A temporary or permanent bank can be used for testing. Sometimes the panel can be designed to allow a set of power supplies from the grid to perform load testing.

2.1.7. Diesel Generator Properties

Automatic transfer switches

The ATS switches installed on the generator motor always monitor the load on the generator motor and, in unusual circumstances, according to previous settings send the appropriate response via electronic switchboard controls and automatically control the entire situation to ensure that the generator motor is in the best condition. To continue. In standby mode, if off-unit power cuts off immediately and with the least time, it will set the ignition time and use of the generator motor.

Automatic voltage regulation

Automatic adjustment of the new voltage causes the motor to function safely and uniformly and to shut off the motor and generator if overloaded, preventing generator failure. Winding of the generator following the latest state of the art technology that prevents any distortion during the work the high-class insulation of the generator prevents any short-circuiting and breakage of the windings, resulting in a uniform and satisfactory working. It provides to the consumer.

2.2. Micro turbine

Micro-turbines are small gas turbines that burn gas and liquid fuels to rotate a turbine that drives an electric generator. The technology of today's micro-turbines is the result of work on automotive gas turbines, emergency power equipment and turbochargers (most of which were launched in the 1950s). Micro turbines entered operational tests in 1997 and from 1999-2000 Business activities began on them. Micro turbines (available on the market or underproduction) range from 30 to 400KW, while conventional gas turbines range from 500KW to more than 300 megawatts. Like larger gas turbines, micro-turbines can only be used to generate electricity or be used in combination to generate electricity and heat simultaneously. Micro turbines can operate with a wide range of different fuels. Generally speaking, micro-turbines have less electrical efficiency than similar piston generator sizes. However, due to its simpler design and relatively smaller moving parts, the micro-turbines have the potential to be easier to install, higher reliability, lower noise, and vibration, require less maintenance and possibly lower capital costs than piston engines generators. Achieving this potential depends on early business successes and increased sales rates to provide more production and sales of service and maintenance infrastructure and technology development.

Micro turbines also have less environmental impact than piston engine generators (e.g. NO emissions and lower carbon monoxide emissions). Gases emitted from micro-turbines are up to eight times less than diesel generators, and the available micro bubbles produce 50% less nitrogen oxide (NO) than natural gas-fired engines. They burn it.

2.2.1. Micro-turbine applications

Due to the flexibility in connection methods, the possibility of parallelization to provide for larger loads, the ability to provide power with less reliability and the lower contamination of micro-turbines can be used for many distributed generation applications. Applications of micro-turbines in a situation where only electricity is generated can include peak demand modification, load-based applications such as overloading, or network support. In the case of co-generation of electricity and heat, the heat from the exhaust gases uses a micro-turbine to provide indoor hot water to heat the building's premises, and this heat can be used to set up drying equipment or absorption cooling as well as to meet other thermal needs in buildings or industrial processes. End-users for these micro-turbine applications include light industrial facilities, financial services, information processing, telecommunications, health, housing, retail, offices, schools, and other industrial or commercial buildings.

Many of the earliest users of the micro-turbine work in the extraction of gas and oil, oil wells, coal mines, whose by-products are gases that provide a free fuel that is usually released into the atmosphere or incinerated incorrectly.

2.2.1.1. Applications in the state of electricity production alone

1. Peak modification

In particular areas, consumers and facilities use dispersed production to reduce costs during pickup. Courier modification can be used for consumers with high electricity consumption costs.

We usually do not have heat recycling in this application, although recycling can be cost-effective if peak periods of consumption are over 2000 hours per year. In general, low equipment costs and high reliability are the primary requirements for this application.

Where courier correction can be combined with other applications such as uninterruptible power supply, this use is substantially strengthened economically.

2. Power supply

Most consumers who need higher reliability or power quality and are willing to pay a higher price for it choose one of a variety of scattered shapes for cost-effectiveness. These customers are less concerned about the cost of power generation equipment than others. In this application, the current high cost of micro-turbines can be justified by their advantages in less pollution, less vibration, ease of installation, availability, reliability and power quality. Particular importance is in the use of noise, vibration and emission power.

3. Power supply in remote areas

In places where grid power is too expensive or unavailable, micro-turbine installation can be a viable option, electricity points in such locations are out of reach, so connect to the local grid by installing a power distribution system. The network is costly to consumers, and the initial cost of installing such a distribution system is usually more than the cost of installing a distributed system. Low emission and long-term fuel efficiency are very important in most off-grid power markets. However, this application can be addressed in the fields of oil production, gas, oil wells, and coal mines, where the byproducts produced are available as cheap fuel. Flexibility in micro-turbine fuel is an important advantage in this application and is evidenced by the early market successes of micro-turbines in this application.

2.2.1.2. Cogeneration applications

Cogeneration systems combine electrical power generation with the use of implicit heat generated. Economically, the production of electric power by micro-turbines will be more cost-effective by the efficient use of heat generated. Heat can be recycled as hot water and low-pressure steam, or hot exhaust gases can be directly used for applications such as heating or drying. The exhaust gas heat can be used to operate heat-activated equipment such as absorption coolers for cooling or drying for dehumidification. Commercial buildings and light industrial installations that require both electrical energy, space heating, hot water, and other heat requirements are usually the best applicants for micro-turbine applications. The simplest thermal requirement is hot water. The primary uses for the simultaneous production of micro-turbines in office buildings and commercial buildings with relatively high demand for electricity and hot water are concurrent, such as universities, hospitals, and residential buildings. The primary uses of cogeneration in the light industry include food processing, chemical, hot water or low-pressure steam forming industries.

2.2.2. Main components and process

Micro-turbines as mentioned above, gas turbines are very small (30 to 400KW), which usually have an internal heat recovery system (called a recuperator) to increase electrical efficiency. In conventional micro-turbines, the inlet air is compressed into a radial (centrifugal compressor) then heated in the recuperator using the heat of the micro-turbine exhaust gases. One or more parts of the turbine expands and provide the torque needed to rotate the compressor and electric generator.

2.2.3. Types of Micro Turbines

Micro-turbines have two types of single-axis and two-axis. In single-axis models, the single turbine rotates both the compressor and the generator. And in dual-axis models, one turbine is used to set up the compressor and another turbine is used to start the generator. The generator turbine exhaust gas is then fed to the recuperator for preheating the air.

Uniaxial models are designed to operate at high speeds (that is, they can exceed 100,000 rpm) and generate high-frequency alternating current. This alternating current is converted to direct current by rectifiers and then converted to alternating current at a frequency of 50 or 60Hz usable to the consumer. The dual-axis micro-turbines have a turbine for compressor operation on one shaft and a separate turbine for generator on another shaft. A generator turbine of this type can operate at a lower speed and higher efficiency the generator turbine is connected to a conventional AC generator with a frequency of 50 or 60Hz by a low-cost single-stage gearbox.

2.3. Photovoltaic

The term Photovoltaic "Photovoltaic" combines the Greek word "Photos" to mean light with "Volt" meaning to generate electricity from light. The discovery of the photovoltaic phenomenon is attributed to the French physicist Edmund Bacvarel, who published his paper on battery life in the year 1839 with the publication of a paper (Becquerel, 1839). He observed that the voltage of the battery did not increase when its silver plates were exposed to sunlight.

In the year 1883, Carlos Edgar Fritz, a New York-based electrical engineer, made a selenium solar cell that was in some ways similar to modern silicon solar cells. The cell consisted of a thin selenium wafer covered with a web of very thin gold wires and a protective sheet of glass. But the cell he built was not very efficient. The efficiency of a solar cell is the percentage of solar energy scattered on its surface that has been converted to electrical energy. Less than 1% of the solar energy emitted to the surface of this primitive cell was converted to electricity. However, selenium cells were eventually widely used in photographic photometers.

2.3.1. Solar cells (photovoltaic)

The key element of photovoltaic technology is solar cells. Photovoltaic cells, commonly known as solar cells, are composed of solid-state semiconductors. Silicon is the most common semiconductor material used by PV cells due to its abundance. Although silicon is an abundant element and makes up a large percentage of the Earth's crust, silicon cells are expensive because of the process of silicon fabrication and purification.

Photovoltaic cells are generated by the use of sunlight and solar cells, generating electricity by dividing the electric pressure is properly constructed semiconductors. Today, the most effective and cheapest solar cell is a material called silica. Sand is one of the most important sources of silica that after crystallization is obtained by the crystallization of silica and then prepared as a plate. In other words, photovoltaic cells, sometimes called solar cells, are made of poles that convert light directly into electricity.

Like transistors, these poles are usually made of thin layers of a semiconductor material such as silicon with small amounts of special additives to create an excess of electrons in one layer and a lack of electrons in the other layer. Photons of light in a single layer generate free electrons, and a conductor string enables the electrons to flow through an outer orbit and access layers without electrons. Photovoltaic panels are made of semiconductors and are formed by P and N-type silicones. When the sun shines on a photovoltaic cell, it emits more energy into the electrons. By sunlight, electrons are polarized in the semiconductor, negative electrons in N-type silicon and positive ions in P-type silicon. This creates a potential difference between the two electrodes, causing the current to flow between them. Because small PV cells are fragile and produce only a small amount of electricity, they form modules. Modules come in a variety of sizes, but they rarely exceed 90 cm wide by 150 cm long for ease of movement. When two cells are connected by a module in a row, their voltage doubles, and when they are connected in parallel, their electrical current doubles.

2.3.2. Photovoltaic system components

The main components of a photovoltaic system are solar panels, storage batteries, AC to AC, controller, steel or building structures, communication cables.

2.3.2.1. Solar panels

They consist of some modules that are interconnected and create arrays of panels. Photovoltaic arrays are generally connected in two series or parallel. These arrays are mounted in either fixed or removable tracers that adapt to the sun's angles according to the season. The detectors are of two types, the detectors rotating on one axis or the two axes, and the detectors always keep the solar panels in the direction of the sun's radiation, thus increasing the output efficiency of the panels by up to 2x.

2.3.2.2. Battery

Usually, a 12-volt battery bank and number of batteries involved and are connected in series and voltage supply system is required. In disconnected systems from the grid, the energy stored in the batteries is used during the night or other urgent situations. Support systems use batteries in the event of a power outage across the network; systems connected to the network do not require batteries.

2.3.2.3. Converter

The electricity produced by the solar panels is converted to DC and converted to AC with the help of converters. The converters are made in different types and sizes and some of them have very high efficiency.

2.3.2.4. Battery Charging Control Device

The battery charging control device is used in photovoltaic systems disconnected from the grid to prevent drainage or excessive battery charging. All standard systems disconnected from the home solar grid have a battery charge control device.

2.3.2.5. Steel or construction structures

They are a major component of photovoltaic systems and hold the module in a particular direction and angle towards the sun. Constructions are made of metal or synthetic material resistant to factors such as wind and rain. Construction structures are designed and selected to suit the position of photovoltaic systems.

2.3.3. Types of Photovoltaic Systems

There are two main types of photovoltaic systems for use in buildings: single and grid-connected. When a power grid connection is not possible or desired, a single system is required. In such cases, it may take several storages to supply electricity at night or on cloudy days and when the maximum amount of electricity is required. The size of the PV arrays is adjusted to inhibit both daytime loads and recharge. In a grid-connected system, a converter is needed to switch the direct current from the PV array to alternating current (AC) at the appropriate grid voltage. It should be noted that in this case there is no need for storage and this will result in significant cost savings and system maintenance. In individual systems, surplus electricity generated during the day is stored in storage for use at night or in dark and cloudy days. Since the price of converters and cells and storage is expensive, a hybrid wind power system is often an ideal complement to the PV system as it not only blows during the night but there is usually significant wind in bad weather as well. Also, in winter, when there is little solar energy to harvest, the air is usually windier than summer. However, not all areas are suitable for wind power use.

۲.۳.۴. Orientation of photovoltaic panels

Maximum solar radiation accumulation occurs when the collector is perpendicular to direct radiation. Since the sun moves both daily and yearly, only a two-way hinge collector can maximize absorption throughout the year. However, hinge collectors can be used only in arid climates, most with direct radiation, and even where 10 to 20 percent of the sun's radiation is scattered. In most sunny and humid climates, about one-second of the sun's radiation is direct, while in the cloudy climate it is 80% or more of the radiation. When integrating with the building, proper orientation and angle should also be considered. The best angle for a PV array is a function of time of year where the maximum amount of power is required. Warm climates require the most electricity during the summer for air conditioning, while cold climates require maximum electricity in the winter and pumps and fans for heating and lighting systems. Optimal orientation is usually south, however, but there is a slight drop in the system up to 20 degrees east or west, although daily loads can affect the orientation.

2.3.5. Photovoltaic integrated building

Photo-voltaic can be used today in existing and new buildings. Its application in building cover is very diverse and opens new avenues for creative designers. Since the photovoltaic cells are the source of sunlight, so the cells are located on the walls of the building which are suitable for direct sunlight. Therefore, the use of photovoltaic miles is often exterior and exterior surfaces of the building. Photovoltaic cells are made in a glass of different colors so that engineers can use them to beautify buildings in addition to their main function. These cells can transmit between eighty and ninety percent of sunlight. This quality enables solar-powered windows to help keep the air cool in the summer and to provide the electricity needed to beautify the building.

2.3.6. Advantages of Photovoltaic Systems

1) The mature photovoltaic technology is robust, reliable and has no moving components and requires little maintenance.

2) It does not require fuel or fuel supply networks.

3) Installing a photovoltaic system is relatively easy and fast, especially for grid-connected systems.

4) Components used in photovoltaic systems have proven their reliability over long periods of use.

5) They are resistant to UV and climate and tolerate high temperatures.

6) They are modular and systems can be of any size.

7) An independent photovoltaic system can supply power to almost any part of the planet.

8) The photovoltaic system reduces greenhouse gas emissions and carbon dioxide emissions.

9) The photovoltaic system generally reduces pollution.

10) The photovoltaic system helps protect scarce resources.

11) Almost everywhere, photovoltaic is a fast-growing market where various businesses can take their place.

12) High shelf life (over 5 years)

2.3.7. Disadvantages of using photovoltaic systems

1) The cost of producing electricity by PV cells is higher than the cost of producing electricity from fossil fuels.

2) Electricity generated from solar energy is unreliable and not always available and the amount of output depends on conditions such as the sun's condition, atmospheric conditions, cloudiness, and so on.

3) The initial cost of installing PV systems is high.

4) Battery storage batteries must be used to use solar energy at night.

5) For high electricity consumption, there is a need for a large area to install PV cells.

6) Lack of specialized and efficient forces to design and install PV systems.

2.3.8. Applications of Photovoltaic Cells

Among the uses of photovoltaic cells are:

1. Supplying Electricity for Electric Fences

2. Provide lighting to remote areas

3. Telecommunication systems

4. Pumping water

5. Water purification systems

6. Electricity supply in rural areas

7. Calculator

8. Watches and toys

9. Emergency systems

10. Vaccine and blood storage refrigerators for remote areas

11. Pool ventilation systems

12. Satellites and Space Equipment.

In general, the applications of photovoltaic cells can be classified into three categories

1. Network Connected Applications

2. Disconnected applications from the network

3. Support Systems Applications

2.3.9. Network Connected Applications of Photovoltaic Systems

The design of grid-connected photovoltaic systems is designed to operate concurrently with the grid. One of the main components of grid-connected photovoltaic systems are converters that convert DC power produced by solar cells to AC voltage and power and automatically cut off power when unnecessary. Overall, there is a bilateral relationship between photovoltaic cells and the power grid so that if the DC power generated by the photovoltaic system exceeds the site requirement, the surplus will be fed to the grid, and at night and in times where climatic conditions allow the use of sunlight. Absent, the electricity required by the site is supplied by the grid. Also, in grid-connected applications, if the photovoltaic system goes out of service for maintenance reasons, the site's electricity will be supplied through the grid.

2.3.10. Disconnected applications from photovoltaic systems network

The design of off-grid systems is independent of the grid and is often designed to generate DC or AC power. Wind power turbines, generators or the entire grid can be used as auxiliary power to generate electricity by off-grid systems, such as photovoltaic hybrids. In disconnected systems, the battery is used to store power and use it at night or when there is insufficient sunlight.

2.3.11. Support Systems

The most important use of photovoltaic support systems during the power outage is the grid. A small photovoltaic support system supplies the power required for equipment such as lighting, computers, telephones, radios, faxes, and larger systems can supply the power required for equipment such as a refrigerator during power outages.

2.3.12. Various solar cell technologies

Photovoltaic systems, which are currently manufactured industrially, are technically classified into two general categories of crystalline silicon as first-generation technology and thin-film technology as second-generation technology. Crystalline silicon cells are divided into mono crystalline, polycrystalline, and striped crystals. Key and important technologies for thin film can also be microcrystalline amorphous silicon (a-Si). The third generation of photovoltaic cells, which are mostly produced at the laboratory level and are being developed by research centers, are referred to as cells with a nominal efficiency greater than thirty-two percent. These types of cells include silicon nanostructures, Up / Down converters, Hot Carrier cells, and thermoelectric cells.

2.3.13. Solar cell production worldwide

The global market for PV cell manufacturing is growing rapidly. This growth has been around 50% in 2003. In 2006, the photovoltaic cell production capacity worldwide reached 2.520MW. This year the proportion of crystalline photovoltaic cells was over 90% and the share of thin-film PV cells was about 8%. Due to the faster growth of thin-film PV cells (about 80% annually), it is predicted that the number of these cells will reach 25% to 30% by 2010.

The number of photovoltaic cells produced in 2007 has reached more than 3.4 GW. Japanese companies, which face a declining share of photovoltaic cells worldwide, have about twenty-six percent of the market. Chinese companies, however, have grown by stunning growth of 20% in 2006 to 35% in 2007.

2.3.14. Installing solar cells worldwide

The installed photovoltaic capacity in the world is growing rapidly. At the end of 2011, the amount was more than 4.67GW equal to 0.5% of global electricity demand. Of this amount, the figure of 27.7 GW is installed alone in 2011, which represents a 67% increase over 2010. In the meantime, Germany alone, with 37% installed capacity worldwide, has reached 24.7 GW at the end of 2011.

Year	Installed Capacity (GW)
2005	5.4
2006	7.0
2007	9.4
2008	15.7
2009	22.9
2010	39.7
2011	67.4

2.3.15. Solar Panel

The solar plate comes from the assembly of solar cells. Since a solar panel generates a limited amount of energy, so the facility consists of several solar panels. Solar panels convert the sun's light energy into electrical energy. Solar panels are made of semiconductor compounds whose task is to convert the sun's light energy into electrical energy. These plates are known as photovoltaic or solar.

Technologically, photovoltaic panels are divided into 4 categories.

• Polycrystalline photovoltaic plates

• Mono crystalline photovoltaic plates

• Thin Film photovoltaic

Everything stimulates the growth of photovoltaic crystals, the use of special coatings or the use of carbon nanotubes to use solar energy cheaper and more efficiently. Recent news suggests that scientists are using technology with a different array structure that is four times more efficient and three times cheaper than current solar cells. The technology was developed at the Royal Melbourne Technology Center and commercialized by a company called Solar Fan. Each solar unit is made up of nine troughs, made of a structure of acrylic and reflective wall lenses that focus solar radiation on photovoltaic cells. This reduces the number of PV cells by 75%. PV cells are used to generate electricity. A temperature transducer is located beneath this section and generates heat for circulating water. There is also a storage tank that is used to store hot water. In addition to maximizing the sun's rays, the array has a motor that directs the sun's rays. The panel says the panel focuses on the global solar power system and can obtain solar energy at a lower cost and greater efficiency than current panels. Each 3.5 cubic meters of this array generates about 2.1 kW of energy. However, standard PV panels with a diameter of about 13 to 14 cubic meters produce the same amount of energy. Solar fan says solar panels can generate about one-fourth the cost of conventional panels with the same energy and heat. This panel model should have ten free samples for testing and be recognized as a different sample worldwide.

2.3.16. What is the difference between a solar cell and a solar panel?

In terms of performance, they are no different. Putting together a number of solar cells creates a solar module. Putting a few solar modules together creates a solar panel, which is generally used for large purposes, with many rows of solar panels forming a series of solar panels.

2.3.17. Types of Solar Systems

Solar systems are divided into two categories: grid-connected and grid-independent. The first batch is used mostly in urban areas and close to the electricity grid. It has the advantage of reducing power consumption and can meet the energy requirement to an acceptable degree during the day. The second category is mostly used in remote areas of the city or away from the electricity grid and is fully independent during the day and night.

Advantages
• No need for an entire network.
• No need for fuel.
• Adaptation to the environment does not pollute the environment.
• No noise pollution.
• It does not require water to generate electricity.

Disadvantages
• The cost of the initial investment is high.
• Dependence on changes in sunlight during different days and months.

2.3.18. Types of Solar Panel Applications
• Independent grid-independent systems
• Power grid-connected systems
• Hybrid systems.

2.3.19 Independent, connected and hybrid systems
• Stand-alone systems: These are systems that are fully powered by solar panels and do not require a full power grid or other power supply.

• Grid-connected systems: These are systems in which electrical energy from solar panels is injected directly into the grid. This system also benefits from the grid while injecting electricity into the entire power grid.

• Hybrid systems: These are systems that use several power supplies to supply the electricity needed and the photovoltaic system is one of the main sources of power. Other sources of energy used in the complex are the grid, diesel generators, wind turbines, etc. (In this model, according to the location and requirement of use of each of the mentioned power sources, they are prioritized and controlled).

2.4. System modeling

2.4.1. Component Modeling (Components)

In this book, the PV deviation angle is optimized by maximizing annual energy production. For this purpose, the measured solar radiation data on a horizontal surface are used to calculate the radiation data on a biased surface. The chosen model used for this calculation is the anisotropy model, while the chosen relation is to calculate the scattered component of solar radiation, the correlation, or the Orgil-Holland relations. This method is usually recommended for surfaces oriented towards the equator [23]. Reference [23] also includes all the equations and models used for this calculation. The genetic algorithm is used to perform this deviation angle optimization, with 1kWp of PV panels selected to calculate their annual energy output. The objective function maximizes the annual energy output of PV panels. Hourly data were used for solar radiation and temperatures. The higher and lower barriers were 0 and 90 °, respectively.

To calculate the generated PV energy, a mathematical model of the PV plate that accurately describes its operation and takes into account the effect of solar radiation and temperature changes must be developed and used. Humidity, wind speed, and other climate indicators also affect the operation of PV panels and their ability to generate power. Their effect is usually indirect, so it is not directly embedded in the mathematical model of the PV plate but is considered in economic technical analysis. The effect of waterborne particles on the surface of sunlight, the effect of moisture on the density and accumulation of dust on PV surfaces, and the effect of moisture on solar cell shielding must be considered. The first effect is considered to affect the solar radiation and the change in solar radiation is also considered in this model. The cost of regular cleaning of the surfaces of the PV panels entails the annual maintenance cost of the PV system for the removal of accumulated dust. The second effect is therefore considered in this analysis. Damaging or reducing the efficiency of PV panels due to their high moisture content for a long time, in addition to other factors, causes the reduction to be considered in the analysis and will be discussed later in this section. Parameters of the PV model such as series resistance (RS), parallel resistance (Rp) and diode desirability factor (a) must be calculated using one of the various methods available to determine their energy use. A mathematical model based on single-diode or two-diode models can be used for this purpose. In the case that PV model parameters are available or calculated, Equation (9) can be used to calculate the PV power generated. In this equation, the voltage of the PV plate and current must be known. But since they are interdependent, a numerical method should be used to calculate the output voltage of the PV plate. If the output current of $I_{PV}(G \cdot T)$ is known. The PV output current can be calculated using (2) where I_{mp-STD} is the maximum power point current under standard test conditions (specified by the manufacturer). This equation can be optimally used for this purpose, while most solar regulators or regulators usually contain a maximum power point tracking circuit that maintains the effective PV output mode at the maximum power point.

$$P_{PV-gen} = V_{PV} * I_{ph} - V_{PV} * I_0 \left[\exp\left(\frac{V_{PV} + I_{PV} * R_S}{a * V_T}\right) - 1 \right] - V_{PV}$$

$$* \left(\frac{V_{PV} + I_{PV} * R_S}{R_P}\right) \tag{1}$$

$$I_{PV}(G \cdot T) = I_{mp-STC} * \left(\frac{G}{G_{STC}}\right) \tag{2}$$

In (2) $G(W/m^2)$ is the calculated radiation on the deflection surface, while G_{STC} is the standard solar radiation test. Is equal to $1000W/m^2$.

In this book, the genetic algorithm is used to calculate the PV model parameter values, which is suitable for the overall range of solar radiation and a wide range of temperatures. For this calculation, the decision vector contains three parameters of the PV model as previously described, but according to the equivalent circuit used to describe the PV model, it may include additional parameters such as saturation or photon current. In the two-diode model, the desirability factor and the saturation current of the secondary diode can also be decided within the vector parameters. In the reference [15] several methods were tested (single or double diodes with saturated and photon fluxes with or without them) and proposed a 3-parameter single diode model due to the fact that it proved the most accurate results.

The lower and higher limits of diode utility constants are 1 and 2, respectively. For the series resistor, the lower and higher resistors are 0.01Ω and 1.2Ω, respectively, while the lower and higher parallel resistors are 50Ω and 1000Ω, respectively.

Briefly, the equations for calculating photon current (I_{ph}), diode saturation current (I_0) and thermal voltage (V_T) are not included here. The equations that take into account the effects of solar radiation and temperature to calculate them are described in detail in [15]. The annual performance level of the PV system decreases by a certain percentage. This decrease in efficiency is mainly due to cellular interconnections and the coefficient of external humidity and consumables in the packaging process. The type of PV panels and their associated manufacturers are the most important factors that affect this reduction. A sample of 1% was proposed by Bourtwin et al. [24] for this reduction. Any loss of mismatch can be attributed to this reduction. This loss expresses the discrepancy between the maximum power of the array as a whole set and the sum of the maximum power of each page. Part of this loss is due to the production defects mentioned earlier, while others are since a scattering of the electrical properties of the coefficients usually occurs when the PV array is composed of more than a few strands [25]. One of the methods followed is to calculate this decrease by offsetting its effect with annual increases in PV system size at 1%, which is equivalent to the percentage determined by the percentage decrease. In this way, the size of the PV system in year n($P_{pv\ sysm\_)n)}$) is added to the size of the PV system in year n-1 and multiplied by the percentage decrease (DPR):

$$P_{pv\ sysm-(n)} = P_{pv\ sysm-(n-1)} + DPR \times P_{pv\ sysm-(n-1)};$$

$$n = 2\cdot3\cdot4\cdot \dots \text{Longevity Project} \tag{3}$$

For other components that make up the hybrid system: Battery bank, charge regulator, diesel generator fuel consumption and two-way inverter, references [3, 18, 26] include detailed descriptions of the models and their respective roles.

Micro-turbine fuel consumption at a given distance depends on the actual power produced at that distance. The manufacturer of the micro-turbine usually displays the data on natural gas flow as a fuel for the different loads in a table or a scheme. In a 30kw micro-turbine, a first-order relationship between natural gas flow (FF_{MT}) at (m^3/h) and the output power of the micro-turbine (P_{MT-mod}) at (kW) in Equation (4) is given. Is. This equation is based on a table in Reference [27] that links the consumption of natural gas fuel and the power output of the micro-turbine.

$$FF_{MT} = 0.314(P_{MT-mod}) + 1.548 \tag{4}$$

Modified micro turbine output power equals micro turbine output power after calculating the effects of ambient and altitude on the sea level on the amount of power produced by the micro-turbine. Equation (5) provides a relation that computes these effects. This equation is developed based on the graph prepared in [27]. The first graph connects the micro-turbine output power and ambient temperature, while the second graph connects the micro-turbine output power to altitude;

$$P_{MT-mod} = P_{MT-rat} * (1 + 0.0001 * h) * (0.911 - 0.005 * T_{amb}) \qquad (5$$

P_{MT-rat} is the measured power of the micro-turbine, h is the height in (m) above sea level, and T_{amb} is at C° ambient temperature.

2.4.2. Economic Modeling

Different types of costs must be taken into account when conducting an economic analysis. These costs include the cost of various components and their installation costs, operating and maintenance costs, and replacement costs. Surplus and money are also considered. Economic analysis uses the life cycle cost in this paper. This is when trying to compare different scenarios being analyzed so you can choose the least cost option. For comparison purposes, the cost of producing one unit of energy is calculated for each scenario.

Economic analysis also needs to define the project life cycle time. It determines its life cycle with the maximum life span of the various components in each project. In the hybrid system considered in this thesis, the maximum lifespan for PV panels is 25 years. For each micro-turbine, the manufacturer or manufacturer usually determines the time (in operating hours) before disassembly and the time (It also specifies the operating hours) before replacing the micro-turbine. Battery life depends more on the number of battery discharge-charge cycles and the amount of charge-discharge depth.

Chapter III

Simulation Method and Optimal Measurement Method

3.1. What is a Genetic Algorithm?

A genetic algorithm is a computer science search technique for finding an approximate solution to optimization and search problems. A genetic algorithm is a special type of evolutionary algorithm that uses biological techniques such as inheritance and mutation. Genetic algorithm, known as one of the random optimization methods, was invented by John Holland in 1967. Later, with the efforts of Goldberg 1982, this method found its place and today, due to its capabilities, it is well-positioned among other methods. Genetic algorithms are usually implemented as a computer simulator in which the population of an abstract sample (chromosomes) of solution candidates of an optimization problem leads to a better solution. Traditionally solutions have been in the form of strings 0 and 1, but are implemented in other ways today. The hypothesis begins with a completely random population and continues for generations. In each generation the capacity of the entire population is evaluated, several individuals are randomly selected from the current generation (based on competencies) and modified (deducted or reconstituted) to form the new generation and transformed to the current generation in the next iteration. For example, if we want to model oil price volatility using external factors and simple linear regression, we will produce the following formula: Oil price at time t = Factor 1 Interest rate at time +t Factor 2 Unemployment at time +t Fix 1. We will then use a criterion to find the best set of coefficients and constants to model oil prices. There are 2 basic points to this method. The first is that the method is linear and the second is that we specify the parameters used instead of searching through the "parameter space".

Using Genetic Algorithm, we set up a formula or scheme, which states something like "Oil prices at time t is a function of maximum 4 variables". Then we will provide data for a set of different variables, perhaps about 20 variables. The genetic algorithm will then be executed, which will search for the best function and variables. The genetic algorithm's work is deceptively simple, very understandable, and substantially the way we believe animals have evolved. The population is considered as possible formulas. The variables that specify each given formula are shown as a set of numbers that make up the person's

DNA. The genetic algorithm engine generates an initial population of formulas. Each person is tested against a set of data and remains the most appropriate (perhaps 10% of the most appropriate); the rest are excluded. The most appropriate individuals are mating (DNA element transfer). And change (random change of DNA elements). It has been observed that over many generations, the genetic algorithm tends to produce more accurate formulas. While neural networks are both nonlinear and non-parametric, the great attraction of genetic algorithms is that the final results are more noticeable. The final formula will be visible to the human user, and conventional formulas can be applied to these formulas to provide a level of confidence. The technology of genetic algorithms is constantly being improved, for example by the equation of viruses that are produced alongside formulas to break the weak formulas and thus make the population stronger overall.

Briefly stated, the genetic algorithm is a programming technique that uses genetic evolution as a problem-solving paradigm. Most of them are selected at random.

Genetic Algorithm is a computer science search technique for finding optimal solution and search problems. Genetic algorithms are one of the types of evolutionary algorithms that are inspired by the science of biology such as inheritance, mutation, natural selection and composition.

Solutions are generally represented as 0 and 1, but there are other display methods. Evolution begins with a completely random set of entities and is repeated in subsequent generations. In each generation, the best are chosen, not the best.

A solution to the problem is shown by a list of parameters called chromosomes or genomes. Chromosomes are generally represented as simple sequences of data, although other data structure types can also be used. Initially, several features are randomly generated to create the first generation. During each generation, each attribute is evaluated and the value of proportion is measured by the function of proportion.

The next step is the creation of the second generation of society, based on selection processes, the production of selected traits with genetic operators: linking chromosomes to each other and changing. For each individual, a pair of parents is selected. Choices are such that the most appropriate elements are selected so that even the weakest elements have the chance of being selected to avoid approaching the local answer. There are several patterns to choose from: roulette, racing selection...

Genetic algorithms usually have a probability of being between 0.6 and 1 indicating the probability of a child being born. Organisms are reunited with this probability. Binding creates two child chromosomes, which are added to the next generation. These are done to find the right candidates for the next generation of answers. The next step is to change the new children. Genetic algorithms have a small, constant change probability, usually of a degree of about 0.01 or less. Based on this probability, the child's chromosomes change or mutate randomly, especially with the mutation of bits in the chromosome of our data structure.

This process creates a new generation of chromosomes, which is different from the previous generation. The whole process is repeated for the next generation, the pairs are selected for the mix, the third generation populations are created, and so on until the last step.

3.1.1. Operators of a Genetic Algorithm

In each case, two elements are needed before a genetic algorithm can be found to find a response to the application: First, a method is needed to provide an answer that the genetic algorithm can operate on. Traditionally, an answer is represented as a string of bits, numbers, or characters. For example, if the problem considers any possible weight for a backpack without breaking the backpack (see the backpack problem), a method of providing an answer can be considered as a string of bits 0 and 1, with 0 or 1 being the plus sign Whether or not weight is in the backpack. The proportion of response is measured by determining the total weight for the proposed answer.

The optimization procedure in the genetic algorithm is based on a random-guided procedure. This method is based on the theory of gradual evolution and Darwin's fundamental ideas. In this method, a set of objective parameters is randomly generated for some constants called populations, after executing a numerical simulator that represents the standard deviation or we fit that set of information to that member of the population. We repeat this procedure for each of the created members, then formulate the genetic algorithm operators, including fertilization, mutation and next-generation selection, and continue this process until the convergence criterion is met.

Commonly, three criteria are considered as a stop criterion: 1. Algorithm execution time 2. The number of generations created 3. Error Criterion Convergence

3.1.2. Applications of Genetic Algorithm

• Hydrological Tracking of Runoff Run off in Dry River Network

• Help solve multi-criteria decision-making problems

• Multipurpose optimization in water resources management

• Optimization and rearrangement of power distribution networks

Termination conditions for genetic algorithms are:

• Get a fixed number of generations.

• Finish the allocated budget (calculation time/money).

• Find an individual (child produced) that meets the minimum (minimum) criterion.

• Get the most out of the children or get better results.

• Manual inspection.

3.2. Simulation method

For every hour of the year, the energy is balanced and a simulation program has been developed for this purpose. The purpose of the program is to simulate a hybrid system that includes more than one energy source following a strategy defined to manage the power flow through the hybrid system. The power flow between different sources in the hybrid system and the priorities that determine this power flow is determined according to this strategy. This strategy is based on simulating the use of the PV system. Therefore, the energy generated by the PV panels and stored in the battery bank is a priority for storing the load. If this energy does not include load conditions, a decision should be made to launch the micro-turbine as the source of the waiting state.

In certain cases where the energy of the PV panels produced exceeds the load and load conditions of the charge, it is used once and temporarily to dispose of this additional energy. As mentioned earlier, a decision to launch a micro-turbine is made when the battery bank becomes discharged to its maximum depth of discharge and there is not enough power generated by the PV system to save this time.

This goes on to recharge the battery completely, with the dual inverter acting as a rectifier and charging the battery.

The energy balance in this study is based on the assumption that no power cuts occur all year (zero without load energy) or that a certain amount of LLP is allowed. The value of LLP is calculated using the following equation.

$$LLP = \frac{\sum_{h=1}^{h=8760} \text{Energy deficit(h)}}{\sum_{h=1}^{h=8760} \text{load demand(h)}} \tag{6}$$

The energy shortage (h), which is the amount of energy needed to load at a given hour, but cannot be covered by the different production of storage resources, can be calculated using the following formula.

$$\begin{aligned} \text{Energy deficit (h)} &= \text{load demand (h)} \\ &\quad - \left[E_{PV}(h) + E_{MT_E}(h) + E_{MT_H}(h)E_B(h-1) \right] \end{aligned} \tag{7}$$

$E_{PV}(h)$ is the energy produced by PV panels at a given hour, $E_{MT_E}(h)$ is the electrical energy produced by the micro-turbine at a given hour. $E_{MT_E}(h)$ The heat produced is a micro-turbine and directly used with heat load $E_B(h-1)$ is the energy stored in the battery at the end of the previous year.

3.3. Optimization of component measurement based on a genetic algorithm

A genetic algorithm is one of the new algorithms that can be used to solve optimization problems in different aspects of life. The individual solutions that make up this population are selected at random at each step. There are three basic commands used with genetic algorithms to form the next population from the current population. Selection commands, initial focus commands, and transformation commands are also formed [28].

As mentioned earlier, the objective is to optimize the component sizes that constitute a new system to be predetermined by the different objective functions. The genetic algorithm has been used to perform this optimization problem and a MATLAB code has been developed for this purpose. This code includes genetic algorithm programming and optimization fit function programming.

In this code, initial population production is formed by random production of population members. (Possible solutions)

Every possible solution is a code of decision vector that considers higher and lower barriers. This first period consists of all successive iterations, and the members of each period are evaluated to calculate the COE proportionality function, which is the minimum COE selected by member processing.

In this method, and after many performances, it is found that a population size of 50% is sufficient to guide this optimization, while the number of courses required for the maximum optimal solution is 80. In most cases, the number of courses required is less than 60. The first focus factor is 0.8, while the transformation factor is 0.2, the focus function is the first mathematical function. The Gaussian transformation function is selected and the stopping criteria is the number of periods specified.

Other types of functions can be used well and this depends on the optimization problem. Whether this is conditioned or not, and the number of parameters optimized, are within the factors that determine the functions to be used.

These functions, which show some results with fewer periods, are functions that have already been defined. Further discussion of these designated functions and other functions can be found in the MATLAB Optimization Toolkit Help List.

For each period, the LLP value is calculated for each population member. This member does not meet the load requirements and a certain amount of LLP is excluded from the population. The first focus, mating, and process of negative change take place on successful members. At the end of the transformation process, each excluded member is replaced by another, maintaining the population size as the period begins. This process is continued until the courses (repetitions) are completed or the stopping criteria are met. In each iteration, the member with the lowest COE is selected.

The parameters that make up the decision vector are optimized: PV plate type (TPV), type of battery (TB) type of micro turbine type (TMT), type of PV installation (TF), PV plate number (NPV) PV Number of batteries (NB), Number Micro turbines (NMT), angle of deviation (β) of PV plates based on COE minimization, and surface angles of (γ) PV plates based on COE minimization. In general, the decision vector (PV) is determined by the following equation:

$$DV = \left| T_{PV} \cdot T_B \cdot T_{MT} \cdot T_F \cdot N_{PV} \cdot N_B \cdot N_{MT} \cdot \beta \cdot \gamma \right| \tag{8}$$

The objective function that is optimized is COE. The goal is to minimize this proportionality function. As mentioned earlier, COE is the cost of generating 1 kWh of energy used with load, and can be calculated as follows:

$$COE = \frac{TAC}{TALE} \tag{9}$$

The TAC is the total annual cost and the TALE is the total annual load energy. When calculating TAC, different types of costs for the different components that make up the hybrid system must be taken into account. Different types of costs have already been mentioned. Also, the gas emission costs, most of which are So_2, No_2, CO_2, were included in the TAC calculation. Tables 3-1 in the subsection below show the different types of costs associated with different components.

The lower and higher conditions that must be considered for the various parameters when the optimized function of the genetic algorithm is solved depend on the parameter itself. For the type, the lower condition must be greater than or equal to zero, and the higher requirement is the number of variants available for the optimization process, while for the lower number, the lower condition must be greater than or equal to zero and, in theory, no restriction should be applied to the higher condition.

3.4. Simulation input

The small remote communities provided by these proposed hybrid systems are located in various locations in the Palestinian Territory. To determine the load curve for one of these distant communities located in the upper region (N 8/31 latitude, E 23 35 longitude) a neighboring electricity community is selected. Surveys and questionnaires were conducted in this community. This community may have the same consumption status as an adjacent non-electric community.

Table 3-1 Specifications of different types of batteries used in the analysis

Battery type				
Specifications	Type 1	Type 2	Type 3	Type 4
Battery capacity (Ah)	2430	1700	1215	648
Battery voltage (V)	2	2	2	2
Battery charge (unit/$)	1271	836	609	305
Battery maintenance cost (year/unit /$)	15	11	8	4

In a hypothetical region, the climate is seasonal, so seasonal consumption changes. Two-class is intended for this load consumption. One of these floors was for the summer (warm) while the other was for the winter (cold). For each floor, there is one working day and one weekend day. Total load measurements for each class were performed for a full week and mean hourly values were obtained. The principal load for the two classes was estimated after analyzing the results of a questionnaire that was prepared for this purpose. This questionnaire included many questions that helped to form the heat load width diagram. These questions determined the type of heat load that was carried out in the community during each hour. Therefore, the heat load directly supplied by the micro-turbine is determined.

As mentioned earlier, hourly solar radiation and ambient temperatures are needed to calculate the solar power produced. In the cases in question, it was observed that the annual mean daily solar radiation on a horizontal surface of the day is $5.94 \, \text{kW h/m}^2$, whereas the hours of sunlight per year are more than 300 h.

For optimization purposes, 4 types of PV panels, 4 types of battery units, 2 types of fittings and 2 types of micro turbines were selected from different manufacturers. As mentioned earlier, determining one type while optimizing the number of units of this type, or optimizing both types and numbers, are possibilities in this study. The various ratings and costs of the selected types of battery units, PV panels and micro turbines are listed in Table

3-3. The other inputs needed for the simulation program are shown in Table 3-4. The amount of CO_2 (in kg/MWh), No_2 and So_2 gas emissions produced by the micro-turbine and diesel generator, and the cost of each of these emissions (in $/ kg), are also included in this table [27].

The first type of fixture is mounted on a roof or ground and may be positioned at an angle of deviation, while the second type is a single-axis follower that can closely follow the sun's motion, this single-axis device moving in a single direction. Maintains the angularity of the surface of the plates to be equal to the solar angularity. The number of pages maintained with each installation depends on the type of installation and the type of pages. To analyze and optimize renewable energy systems in this book, a MATLAB-based software has been developed that optimizes an off-grid hybrid system based on PV as a renewable resource. The optimization performed in this software involves optimizing the angle of deflection, optimizing the component sizes, and also the type (brand name) of these components. The type of accessories for PV panels is based on optimized parameters.

Table 3-2 Specifications of Different Types of PV Plates and Similar Installation Supplies Used in the Analysis

PV type				
Specifications	Type 1	Type 2	Type 3	Type 4
Rating (W)	135	175	220	235
Cost ($/W)	2.23	1.37	1.54	1.18
Installation cost ($/W)	0.7	0.7	0.7	0.7
Maintenance Cost (Year/$ /W)	0.025	0.025	0.025	0.025
Number of PV panels in type 1 installation structure	2	2	2	2
Number of PV panels in type 2 installation structure	11	12	9	9
Type 1 installation cost (unit/$)	263	289	397	397
Type 2 installation cost (unit/$)	2452	3636	3636	3636

Table 3-3 Specifications of the two types of micro turbines used in the analysis

Micro turbine type		
Specifications	Type 1	Type 2
Rating (kW)	30	65
Investment Cost ($/kW)	2970	2490
Maintenance cost (year/$/kW h)	0.02	0.0175

Table 3-4 Other inputs of the simulation program

Item	Amount
Bilateral inverter investment cost ($/kW)	750.0
Investment Cost of PV Solar Charger Converter ($/kW)	450.0
Diesel Generator Investment Cost ($/kW)	550.0
Two way inverter efficiency (%)	92.0
PV Solar Charger Converter Efficiency (%)	95.0
WH battery units efficiency (%)	85.0
Lifetime of PV panels (years)	25.0
Battery life (years)	6.0
Project Lifetime (Year)	25.0
Price of natural gas fuel ($/m^3)	0.33
Price of diesel ($)	1.77
Fuel filters and air handling time (h)	8.0000
Replacement time of thermocouples, lighters and fuel injectors (h)	16.000-20.000
Micro Turbine Battery Life (h)	20.000
Overhaul time (including replacement of core turbine) (h)	40.000
Turbine displacement time (h)	80.000
Diesel engine displacement time (h)	24.000
Interest rate (discount) (%)	8.0
General Inflation Rate (%)	4.0
Fuel Inflation Rate (%)	5.0
CO_2 emissions from micro turbine (kg/MW h)	787.4
Production of NOx gas from micro turbine (kg/MW h)	0.245
Generation of SOx gas from micro turbine (kg/MW h)	0.008
CO_2 emissions from diesel engine (kg/MW h)	649.5
Generation of NOx gas from diesel engine (kg/MW h)	9.89
Generation of SOx gas from diesel engine (kg/MW h)	0.206
CO_2 cost ($/kg)	0.014
NOx cost ($/kg)	4.2
SOx cost ($/kg)	0.99

Chapter IV

Simulation Results and Analysis

4.1. Mathematical Modeling Results of PV Plates

The method used for modeling PV plates depends on the extraction of parameters using a genetic algorithm. Table 4-1 describes the values of the 3 parameters used in the model for each PV plate type as shown in table 3-2. The percentages of the mean absolute errors in the currents are calculated by showing them in a similar flow at the maximum power point. The calculated error is the difference between the PV output current calculated using the extracted parameter values and the current generated by the manufacturer.

4.2. Solar Data Analysis Results (Optimizing Screen Angles)

The genetic algorithm has been used to optimize PV deflection angles and surface-oriented angles, and when optimized based on maximizing annual PV energy production, a value of 30° has been obtained for PV deflection angles. While directing the PV plates directly to the south in the Northern Hemisphere is usually directed at a certain angle to the east or west of the south, it may lead to more annual energy production. In this study, the optimum value for this angle was + 16° (ie southwest). The fact is that optimum surface angularity is considered when maximizing the annual energy output of PV panels may not be similar to the optimum angle, although important as COE minimization such as this.

4.3. Optimization results of hybrid system measurement

In this book, different scenarios and cases are analyzed to select the most optimal formulation that covers the demand load with the specified LLP value, with minimum COE output.

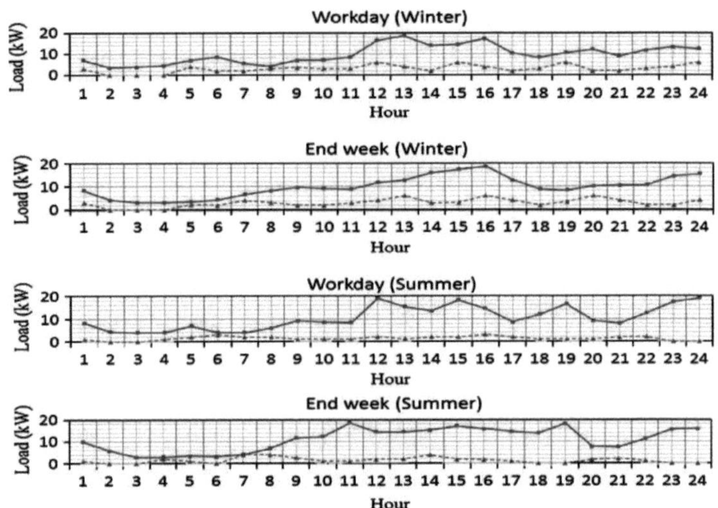

Figure 4-1- Hourly sample diagram (solid red line for total load and blue line for heat load).

Table 4-1- PV Model Parameters (Extracted Using Genetic Algorithm)

PV type				
Parameter	Type 1	Type 2	Type 3	Type 4
Desirability factor	1.175	1.014	1.08	1.037
Series resistance	0.228	0.715	0.371	0.338
Parallel resistance	291.9	987.6	972.6	935.9
Average percentage of absolute error	0.18	0.25	0.19	0.19

4.3.1. Best Optimized Results

Case 1 (Best Optimized): Optimize the component type and the number of similarities of each type. In this case, the deflection and angle-oriented angles are optimized, minimizing COE relative to the optimized angles based on maximum annual PV energy production.

From the previous subsection, the optimum deflection angle that maximizes the energy produced by PV panels is set to 30 degrees, while the optimum surface angle that maximizes energy production is + 16 degrees and ensures that the optimum deflection angle or Surface angularity that maximizes annual energy production does not mean the optimal angle that minimizes COE. In fact, the difference between the calculated COE values for the optimal deviation and oriented angles is not low and calculates the COE minimization and calculates the COE for both the optimized deviation and oriented angles, taking into account the maximum annual energy production.

In this case, the micro-turbine operates in a cogeneration mode, where the heat load is directly supplied by the heat generated during the operation of the micro-turbine (through a heat exchanger), while the micro-turbines electrical output provides only the electric charge. In this case, the micro-turbine operates at its rated power to provide electric charge and the surplus generated power is used to charge the battery inside the dome inverter. In this case, the micro-turbine electrical output provides both electrical and heat loads, which is case 14, while in the case where the micro-turbine operates according to load changes, it is case 15. In this case, there is no surplus power available to charge the batteries.

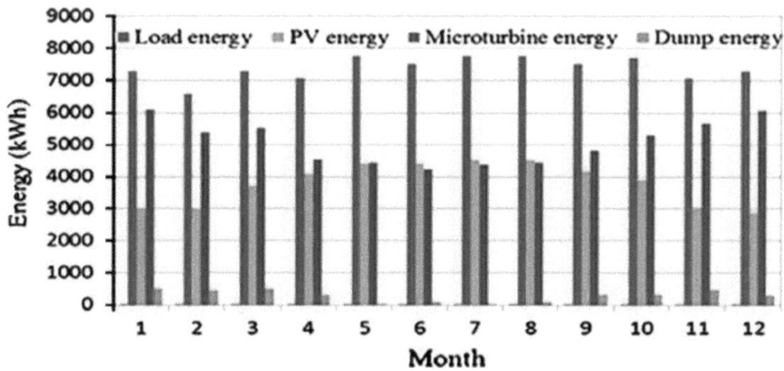

Figure 4-2- Energy generated by PV plates, micro turbine generation energy and discharge energy monthly

Table 4-2- Simulation results of different cases

case	PV type	Battery type	Micro turbine type	Type of PV fixture	Number of PV panels	Number of battery units	Number of micro turbines	Angle of deviation	The azimuth angle of the surface	COE ($/kWh)	Turbine Working Hours (h)	LLP (%)
1	4	4	1	1	92	20	1	30	14	0.2587	2692	0
2	4	4	1	1	92	20	1	32.8	14	0.2589	2696	0
3	4	4	1	1	92	20	1	30	16	0.2588	2694	0
4	4	4	1	1	92	20	1	30	0	0.2595	2711	0
5	4	4	1	1	92	20	1	32.8	0	0.2596	2713	0
6	4	4	1	2	80	19	1	36	-	0.2662	2778	0
7	2	2	1	1	117	8	1	29	14	0.2627	2764	0
8	4	2	1	1	87	8	1	29	17	02601	2760	0
9	2	4	1	1	120	20	1	25	15	0.2611	2726	0
10	1	1	1	1	149	6	1	30	14	0.2887	2780	0
11	3	3	1	1	92	11	1	30	12	0.2659	2777	0
12	1	3	1	1	150	12	1	29	12	0.2783	2774	0
13	1	4	1	1	137	23	1	30	14	0.2751	2865	0
14	4	4	1	1	88	19	1	29	14	0.2657	2921	0
15	4	4	1	1	80	12	1	29	16	0.3037	6366	0

16	-	-	1	-	0	0	2	-	-	0.3273	8760	0
17	-	-	1	-	0	0	2	-	-	0.3517	8760	0
18	4	4	-	1	490	337	0	40	13	0.5140	-	1.99
19	4	4	-	1	507	416	0	30	14	0.5752	-	0
20	4	4	-	1	459	302	0	30	15	0.4739	-	4.95
21	4	4	1	1	230	129	1	32	15	0.3982	624	0

In cases 2 to 5, different values of the angle of deviation and surface orientation are considered to calculate their effect on COE.

Case 2: Similar to case 1, but the deflection angle is optimized based on the maximum energy produced annually from the PV system and optimizes the deflection angle to minimize the COE.

Case 3: Similar to case 1, but the angle of deviation is optimized and COE minimization is considered. And surface angularity is optimized to maximize annual PV energy production.

Case 4: Similar to case 1, but the angle of deflection is optimized and COE minimization is considered and the surface angular is southward.

Case 5: Similar to Case 1, but the angle of deviation is optimized based on maximizing the annual generated energy from the PV system and is surface angular to the south.

It is noteworthy that the difference between COE values for cases 1 to 5 is very small. Therefore, the optimized deflection angle based on maximizing the energy generated annually from the PV system and the surface angularity to the south can be effectively used without the need for further mathematical calculations while calculating these angles and providing another basis for evaluation.

4.3.2. Evaluate more than the original optimized

For case 6, the number of PV panels required is lower and, as this type of appliance, the amount of annual PV energy produced is greater than that of type 1. This case follows the motion of the sun in such a way that the angle of the surface is equal to the angle of the sun, which increases the amount of energy produced.

Cases 4 to 13 are selective cases where the types of components or major components are determined and optimization is performed based on the number of components.

Case 14 is similar to case 1, but the micro-turbine electrical output provides both electric and thermal loads. It can be seen that the use of the composite micro-turbine production feature will reduce the COE.

Case 15 is similar to case 1, but the micro-turbine operates in such a way that its electrical output is subject to changes in the electrical load.

In this case, the number of annual operating hours is more than twice the annual operating hours of Case 1, for Case 1, the annual consumption of natural gas is 32.029 cubic meters, while this figure is 30.244 cubic meters. This is due to the fact that the consumption of micro-turbine natural gas depends on the load connected to it.

In the case of 16, the only micro-turbine scenario is that the number of presumed micro-turbines is double the load and the electrical load is the only part that is stored (i.e. the use of the cogeneration feature). In this case, each micro-turbine works half-time. The annual fuel consumption of the first micro-turbine was 18.102 m^3, while the second turbine consumption was 27.252 m^3.

Case 17 is similar to case 16 the temperature of the two turbines here stores the load (electric and heat loads). In this case, the annual consumption of natural gas was the first turbine of 21.843 m^3, while the second turbine consumed 32.382 m^3.

In cases 18 to 20, the load demand is covered by PV and battery (i.e. no micro turbine). The effect of different values of LLP on COE is evaluated in these cases.

Case 18 is assumed to be a standalone system (PV and battery without micro turbine). A certain value for LLP (2%) is assumed to be higher for this case than for the COE calculated for the optimized hybrid as stated in Case 1.

Case 19 is similar to case 18, but reliability is 100% (i.e. zero load acceptance). To accommodate these conditions, a large number of PV panels and battery units are needed to increase the COE value, an increase in COE compared to Case 18 is about 12%. This occurs when critical and critical loads are required (without interrupting power storage) and are located off the grid. Medical clinics and telecommunications stations that are located in remote areas without access to natural gas are one example of this.

Case 20 is similar to the previous two, but an LLP value of 5%. COE is lower than the previous two cases. COE reduction is about 76% compared to case 19 (zero load rejection) In case 21 a diesel generator is considered as a waiting source instead of a micro turbine. When the price of diesel in the Palestinian territories is high. The number of PV panels and battery units increased compared to case 1. The micro-turbine acts as a waiting source. In this case, the number of diesel generators operating hours per year is less than 25% of the total hours of micro turbine operation. The increase in COE for this case is compared with case 1. Which is about 54%. Although the number of micro turbines operating hours is four times higher than that of a diesel generator when used as a waiting source, the annual cost of emitting gas is lower. In a micro turbine, it is about 35.58 per year.

4.4. Designing the optimal state of the hybrid system

As shown in Table 4-1, the number of PV panels for the optimal case is 99, while the number of battery units is 20 units. The recommended DC path voltage is 48 V for this system. The measured voltage is the optimized type battery 2V, so 24 units of battery are needed for serial connection to provide the proposed DC path voltage. That is, four additional units are required to raise the COE to 0.26248 kWh/ $. The optimized PV plate type has a maximum power point voltage of 29.8V. That is, 10 PV panels of this type must be connected in series to form a string while the optimal number of PV panels is 92, and the strings should be connected in parallel. In this case, 90 panels are required for this installation. The COE calculated for 24 battery units (instead of 20) and 90 panels (instead of 92) is 0.26178 kWh/ $.

The measured output voltage of the charger regulator shall be 48 V, while the input voltage shall bear the voltage of the open circuit of the PV array. (369V = 36.9V * 10) Charger regulator power measurement shall not be less than the maximum power of the PV system and 25 kW shall be selected. For these load regulator sizes, a maximum power embedded in the circuit is recommended in their design.

4.5. Optimization effect

The feasibility of using a micro-turbine as a waiting source in hybrid power systems can be generalized. This conclusion is drawn from the fact that the feasibility of using a diesel generator as a waiting source in the hybrid power system, especially for other applications, has been substantiated by previous research and the possibility of using micro turbines as waiting sources instead of diesel generators for this purpose. The book's research is proven.

In this book, the deflection and angle-oriented angles are optimized, minimizing COE relative to the optimized angles based on the maximum annual PV energy output. From the previous subsection, the optimum deflection angle that maximizes the energy produced by the PV panels is °30, while the optimum surface angle that maximizes energy production is + 16 °, ensuring that the optimum deflection angle or angle is maximized. The surface orientation that maximizes annual energy production does not mean the optimal angle that minimizes COE. In fact, the difference between the calculated COE values for the optimal deviation and oriented angles is not low and calculates the COE minimization and calculates the COE for both the optimized deviation and oriented angles, taking into account the maximum annual energy production.

In this case, the micro-turbine operates in a cogeneration mode, where the heat load is directly supplied by the heat generated during the operation of the micro-turbine (through a heat exchanger), while the micro-turbine's electrical output provides only the electric charge. In this case, the micro-turbine operates at its rated power to provide electric charge and the surplus generated power is used to charge the battery inside the dome inverter. In this case, the micro-turbine electrical output provides both electrical and heat loads, which is case 14, while in the case where the micro turbine operates according to load changes, it is case 15. In this case, there is no surplus power available to charge the batteries.

As shown in Table 4-1, the micro-turbine operates around 2692 h annually. That is every day a certain number of hours work. The simulation results show that the energy produced by the micro turbine is about 57% of the total energy produced by the micro turbine itself and the PV plate. These simulation results show that 4% of the total energy produced is discarded, while the total energy loss on an annual basis is about 14%. This means that the energy efficiency of this hybrid system is about 86%.

The cost of micro turbine capital is the highest among other component capital costs. About 49% of the initial costs are total, while about 41% of the total costs are for the PV system, in addition to the PV pages including the PV regulator and battery bank. The rest of the cost goes to the friendship inverter. The energy divided by the micro turbine and PV system for each month is shown in Figure 4-2 the monthly charge and dissipation energies are also shown in the same figure. It is noteworthy that the energy produced by the micro-turbine is higher in the months with less PV energy. During these months, the waste energy is even higher. This is because the micro-turbine operates at its rated power while the heat consumed by the heat exchanger from the micro-turbine directly supplies the load heat energy. In the winter, the micro turbines work long hours, in line with the higher heat load directly supplied by the micro turbines. In fact, this explains more waste energy during the winter months.

4.6. Genetic Algorithm Results

Best Tilt Angle= 29.6972

Best Surface Azimuth Angle= 15.9406

PV Type= 4

Battery Type= 4

Micro turbine Type= 1

PV Fixture Type= 1

Number of PV Panels= 108

Number of Battery Units= 19

Number of Micro turbines= 1

Sun hours of Year= 4655

Running hours of Micro turbine= 2۶۹۲

Battery Size Power= 20.9304kW

Total Year Radiation on The Horizontal Surface = 2083.397kW/m^2

Maximum Radiation on The Tilted Surface With Yearly Optimization= 2255.2457kW/m^2

Yearly Radiation Optimization Percent = 8.2485%

Minimum of COE= 0.26299 $/kW

Minimum of LLP= 0

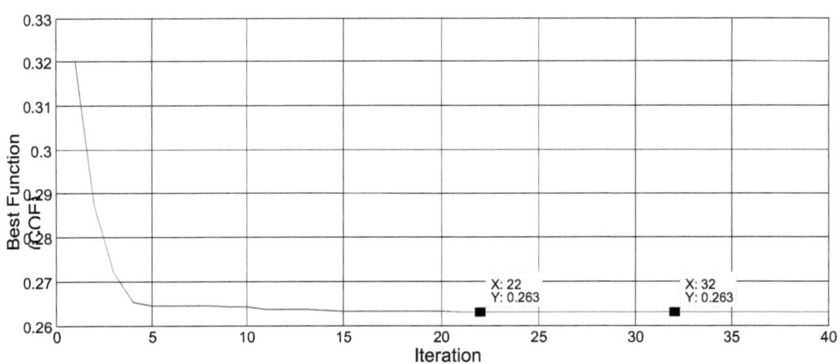

Figure 4-3- Optimization of energy cost function using algorithm function

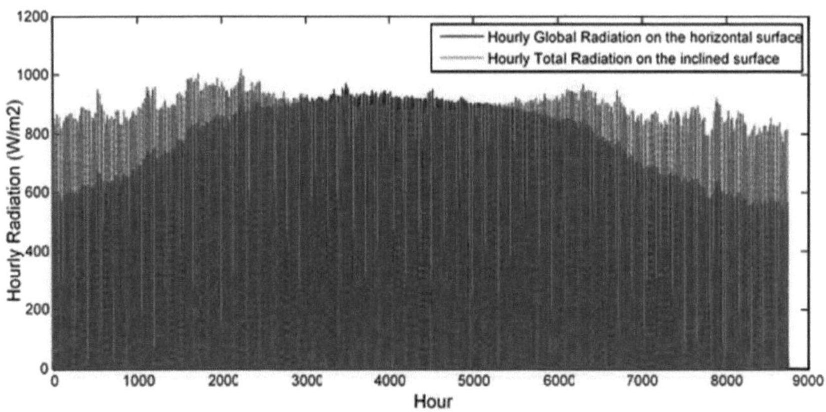

Figure 4-4- Optimization of slope angle and surface angle to increase the amount of annual radiation on the surface of the solar panels

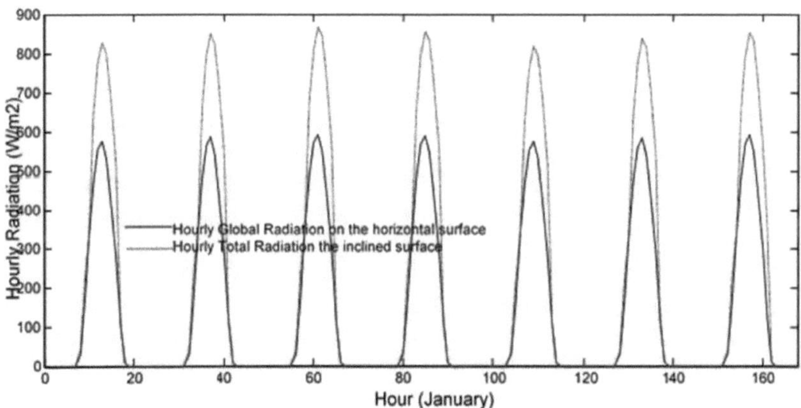

Figure 4-5- Optimization of slope angle and surface angle to increase the amount of annual radiation on the surface of solar panels in January

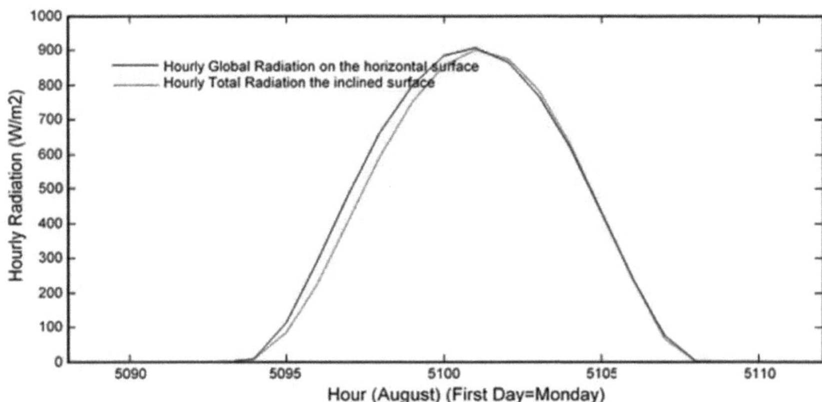

Figure 4-6- Optimization of slope angle and surface angle to increase the amount of annual radiation on the surface of solar panels on the first day of August

This diagram shows the hourly optimization of the horizon and slope (year-round) optimized. The amount of radiation on the ramp is higher than the horizontal, and the amount of radiation is higher than the horizontal. In summer, the sun shines more on the horizon than on the slope. The blue diagram is on the horizontal surface and the red diagram is on the surface we optimized. In the next three seasons, the values are optimized as the radiation is applied to the tilting horizon.

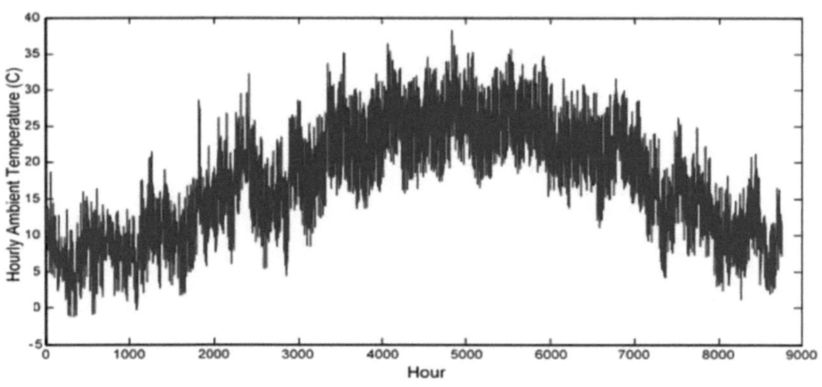

Figure 4-7- Hourly ambient temperature in one year

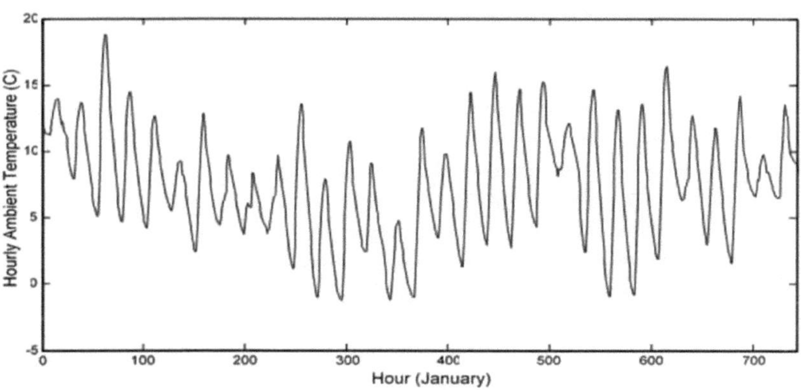

Figure 4-8- Hourly ambient temperature in one year in January

This figure shows the ambient temperature for all clocks throughout the year, according to which the maximum output power will depend on both the amount of radiation and the temperature. The higher the radiation and the lower the temperature, the higher the output power.

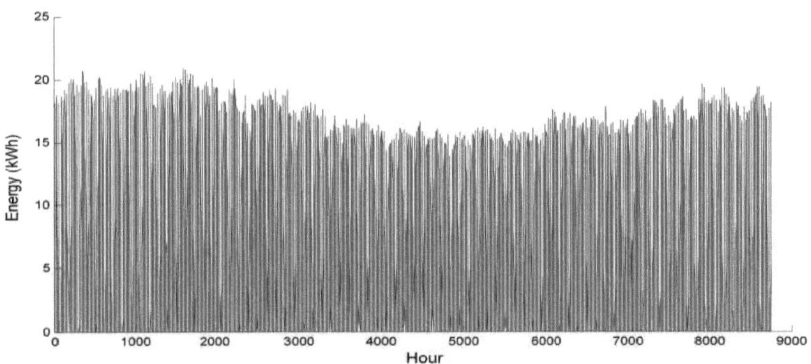

Figure 4-9- Generating energy of solar panels

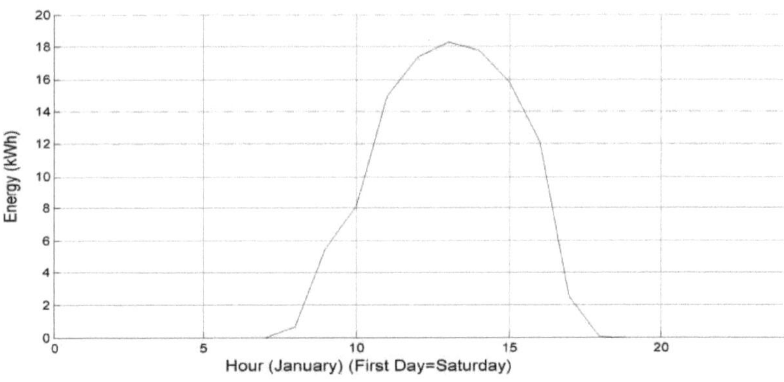

Figure -4-10 Energy generated by solar panels on the first day of January

This figure shows the energy produced by solar panels (solar cell power). According to this chart, we will not be able to have solar panels during the night because we do not have radiation at night. But most solar panels are powered from noon to sunset. During the day when the solar panels are switched on, they both charge and charge the battery.

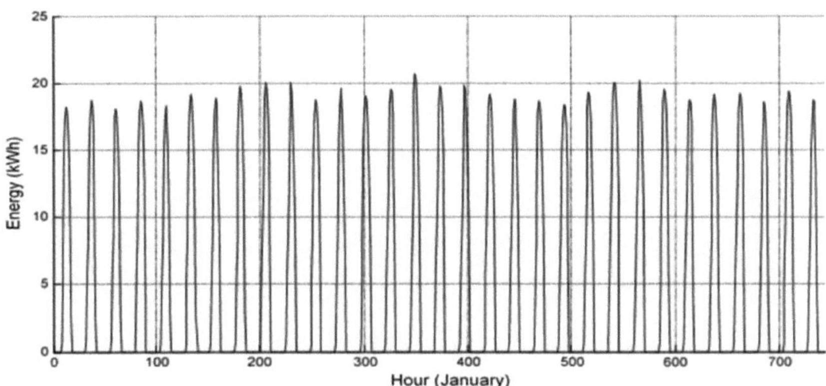

Figure -4-11 Solar panels power generation in January

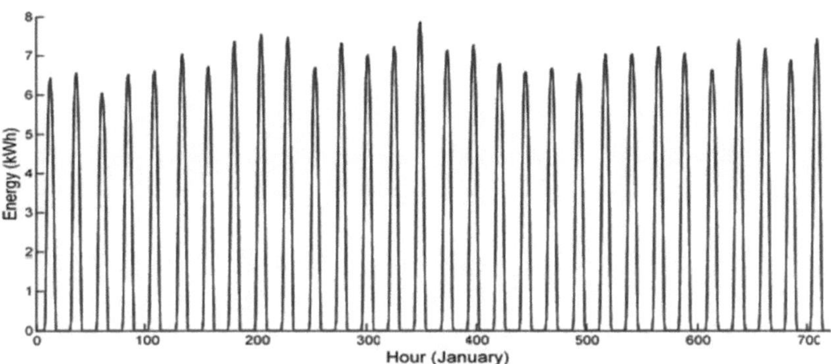

Figure 4-12- Energy generated by solar panels in optimum state

By comparing the two figures above, we conclude that the energy generated by solar panels is at an optimum of about 20 kWh in January and suboptimal is less than 8 kW in January. That is, we will have about 60% optimization of solar energy through solar panels.

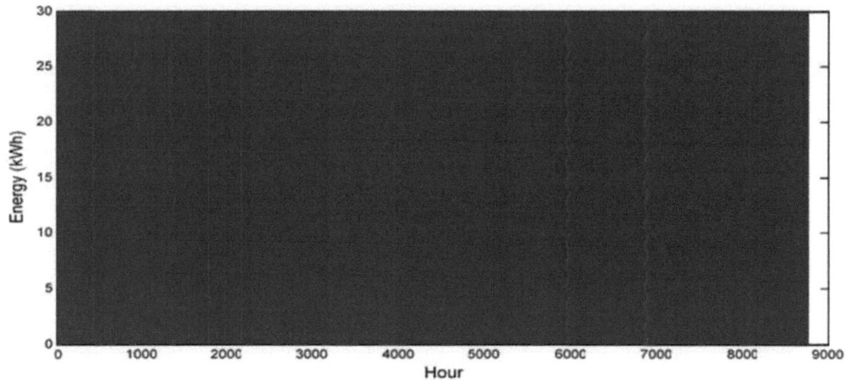

Figure-4-13 Micro turbine Power Generation

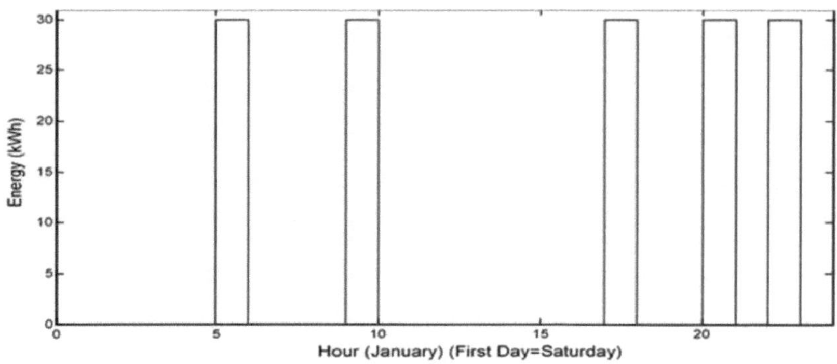

Figure-4-14 Micro turbine energy production on the first day of January

These figures show the energy produced by the micro turbine in KWh. Normally the micro turbine does not stay on much during the day when the sun is shining, but during the night the micro turbine will be on and we can use micro turbine and battery energy at the same time. We use battery power when the battery is charging, and it charges both the battery and the battery when charging. Try to keep the micro turbine as low as possible because of the high cost of micro turbine fuel consumption. The micro turbine delivers 30kW.

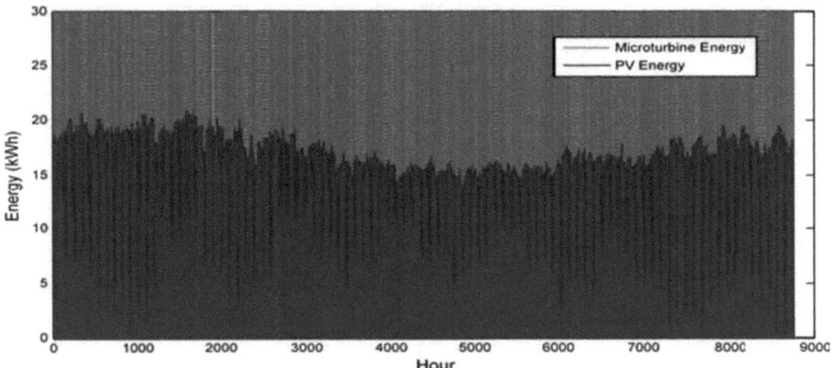

Figure-4-15 Micro turbine and solar panel power generation

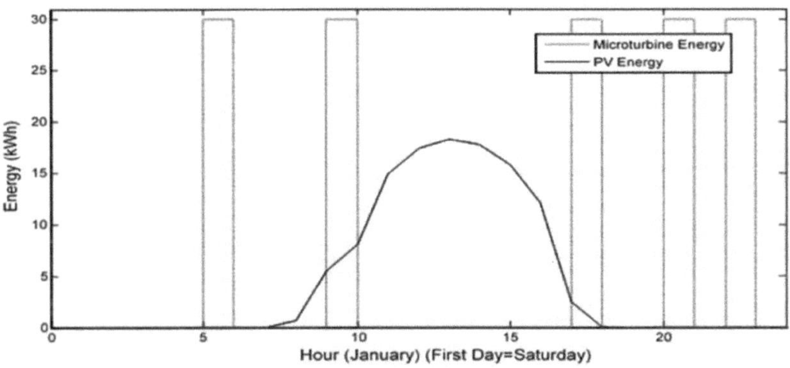

Figure-4-16 Micro turbine and solar panel power generation on the first day of January

These figures show the energy generated by the micro turbine and solar panels simultaneously. During the time we have the sun, the micro turbine does not stay very light. During the night the micro turbine lights up and uses micro turbine and battery energy together. When the battery is charged, the battery is used, and when the battery is charged, the micro turbine turns on and supplies both the charge and the battery.

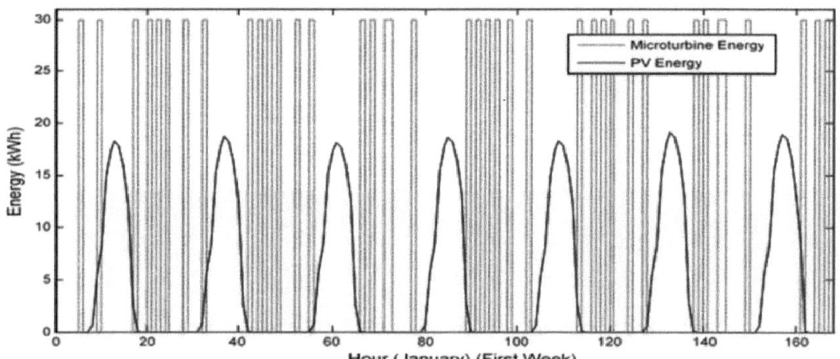

Figure-4-17 Optimum Micro turbine and Solar Panel Power Generation for the First
Week of January

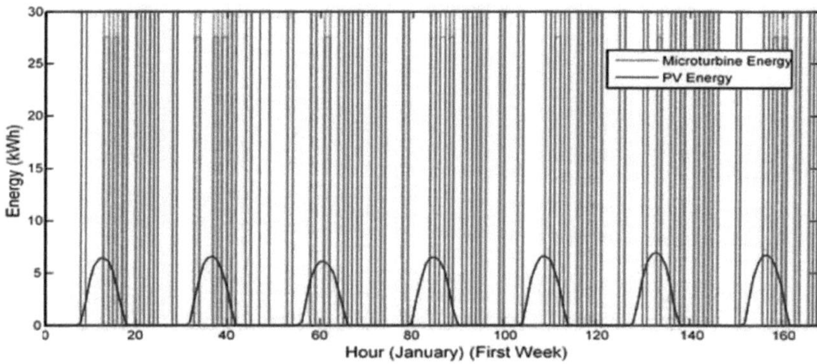

Figure-4-18 Micro turbine and solar panel power generation in the first week of January
in optimum condition

By comparing the two figures above, we conclude that in the optimum state we obtain more energy from the solar panels than in the optimum state. It will, therefore, be less energy-consuming because the solar panel is less fuel-efficient than the micro turbine. In an optimum state, the micro turbine is switched on and off more often, working longer, consuming more fuel and costing more.

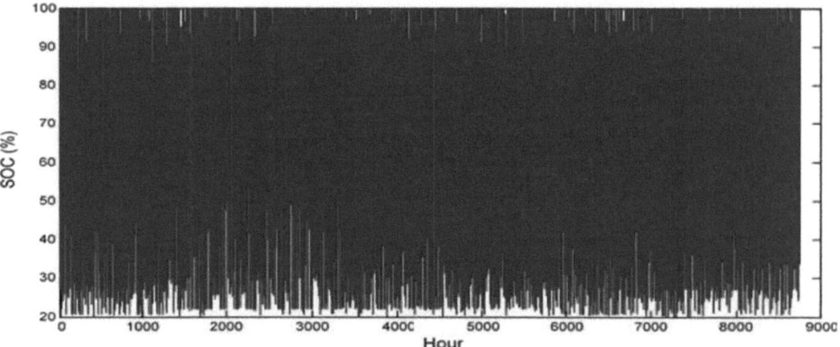

Figure-4-19 Battery charge status percentage

These figures show the average battery charge per month, which is 60-70%. Battery is not used when the weather is sunny as the solar panel provides the load. Battery life is about 6 years. The number of batteries is 19 and is considered to be of type 4.

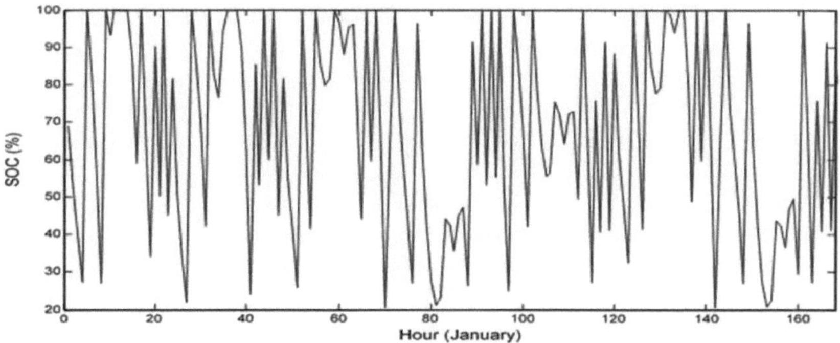

Figure-4-20 Battery status percentage optimized for January

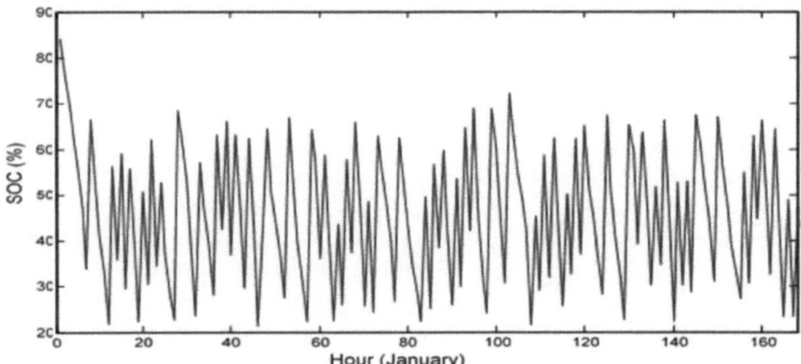

Figure-4-21 Battery status percentage in January at optimum state

By comparing the two figures above, we conclude that in the optimum mode the battery will charge about 80% and in the non-optimal mode about 60%.

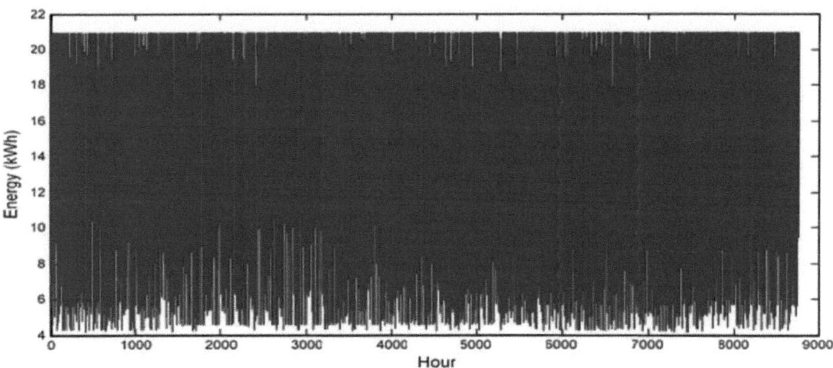

Figure-4-22 Battery power

In this book, the number of batteries is 19 and type 4. The amount of energy the batteries get is obtained by multiplying the battery voltage at the battery capacity (amps). This is about 20 kWh. The minimum battery charge is also 20% because if it is less then the inverter will remove the battery from the circuit.

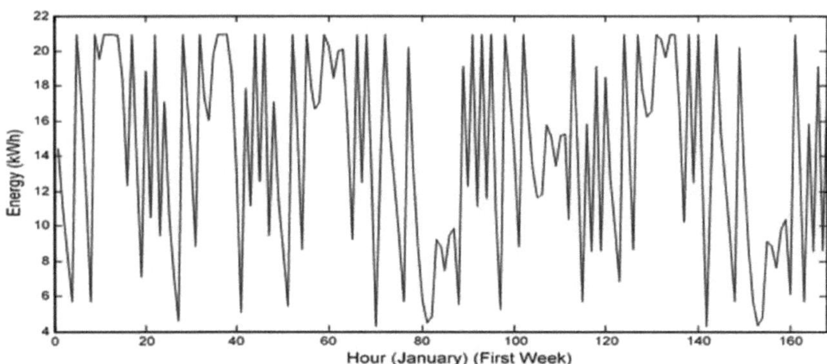

Figure-4-23 Battery power in the first week of January at optimum

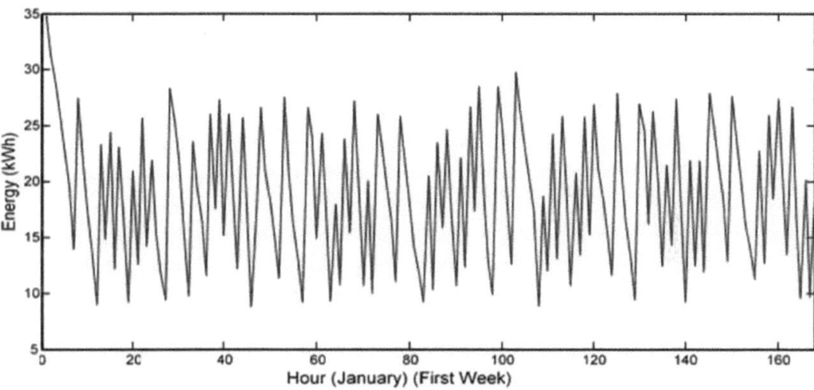

Figure-4-24 Battery power in the first week of January in optimum condition

Optimally, fewer batteries are used, and the batteries are more likely to be recharged, and the battery will be less charged and discharged, resulting in longer battery life.

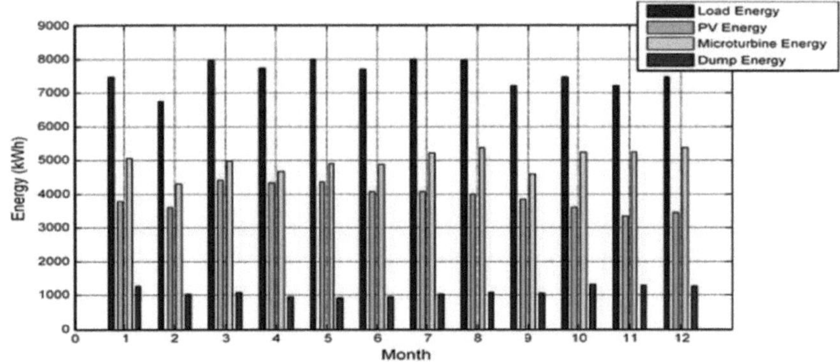

Figure-4-25 Solar Panel Energy, Micro turbine Energy, Load Energy and Discharge
Energy at Optimal State

This figure shows the energy produced by micro turbine and PV and the fine energy. We
will waste energy when the energy produced is greater than the energy consumed. In the
winter, when the micro turbine works longer hours to produce heat energy, we will have
more wastage energy.

Energy Saving = (Solar Panel Energy + Energy Micro turbine) - Energy Consumption

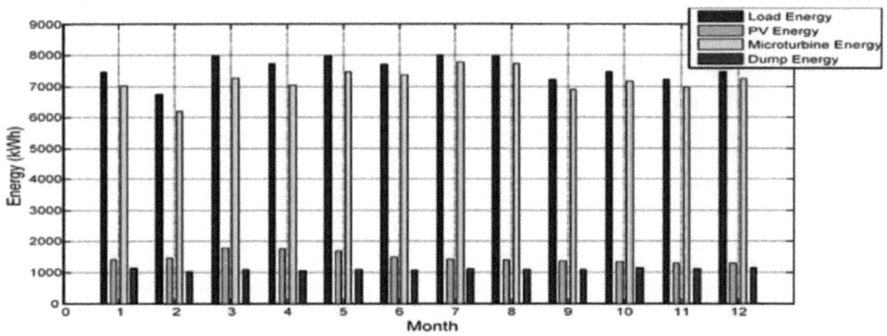

Figure-4-26 Solar Panel Energy, Micro turbine Energy, Load Energy and Discharge
Energy in Optimal Mode

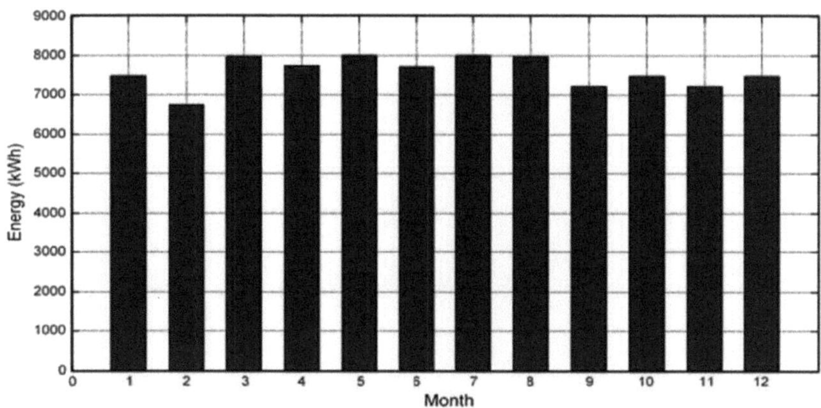

Figure-4-27 Monthly load energy

This chart shows the amount of monthly load energy, which is the highest amount in the middle of the year, at about 8000 KWh.

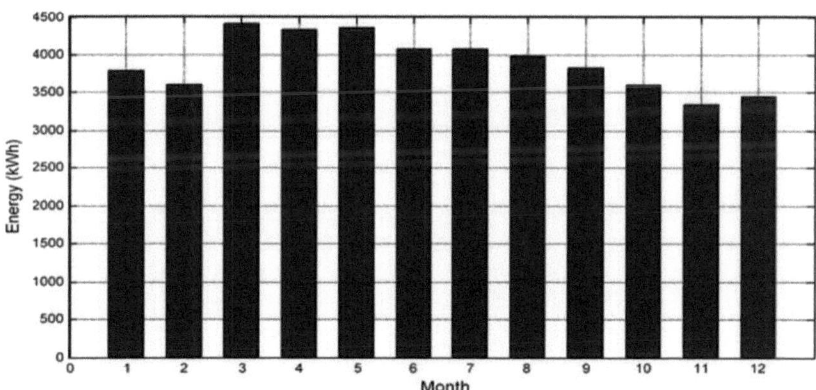

Figure-4-28 Optimum Solar Panel Energy Generated Monthly

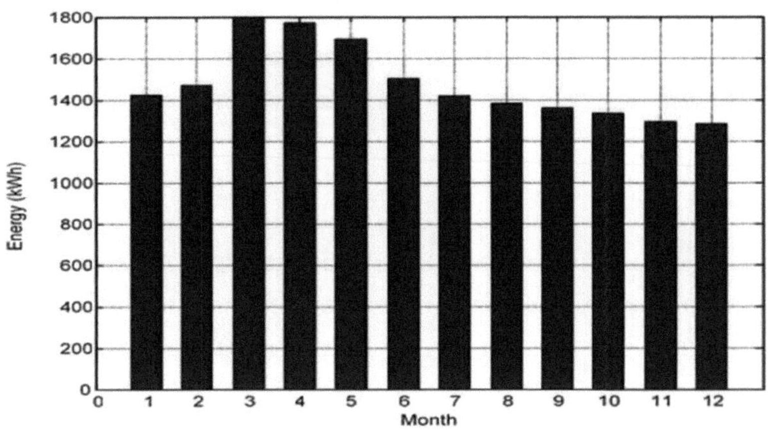

Figure-4-29 Monthly Energy Generation of Solar Panels in Optimal Mode

This figure shows the amount of energy produced by the solar panels in the months of the year, comparing it to the highest amount in the warmer months of 3, 4, and 5. This is about 4500 KWh in the optimized state and 1700 KWh in the optimized state.

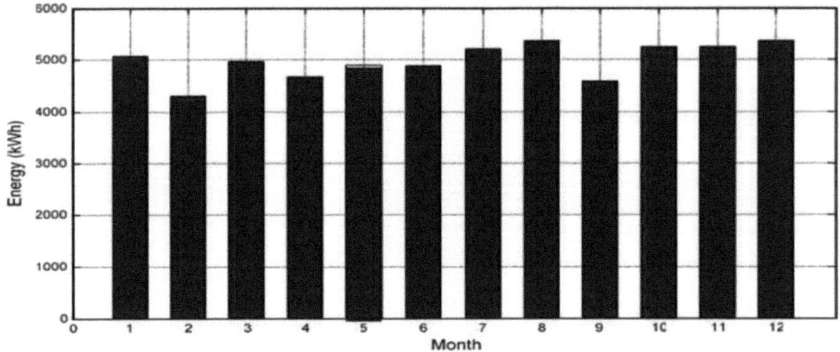

Figure-4-30 Optimal Micro turbine Energy Generated Monthly

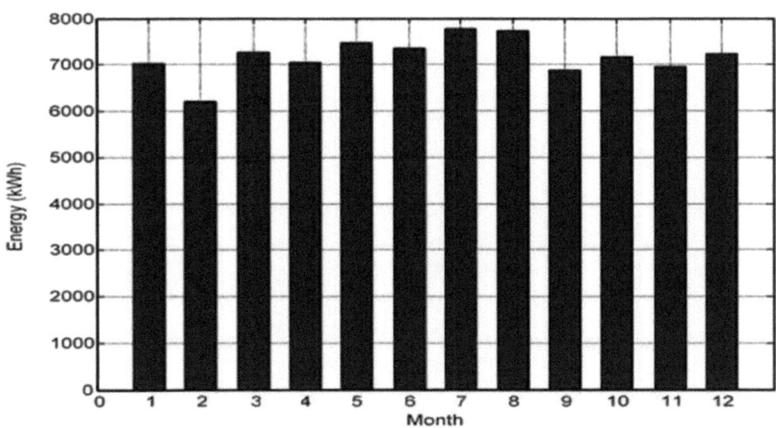

Figure-4-31 Monthly Micro turbine Generation Energy in Optimal Mode

This figure shows the energy produced by the micro turbine in the months of the year, comparing it to the fact that in an optimum state the micro turbine will work longer and generate more energy so we will have more fuel costs.

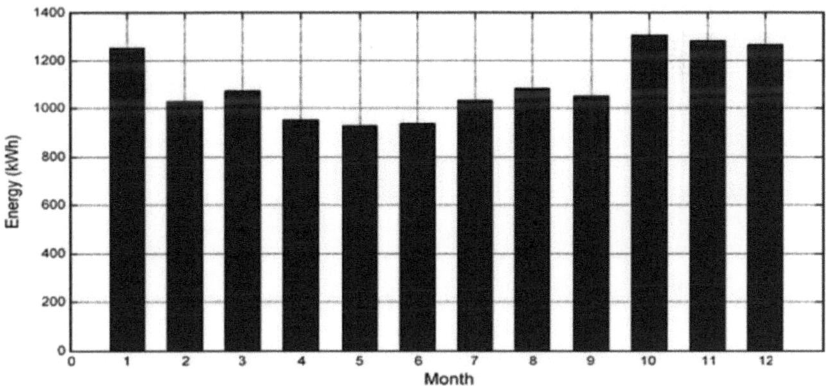

Figure-4-32 Monthly disposal energy at optimum

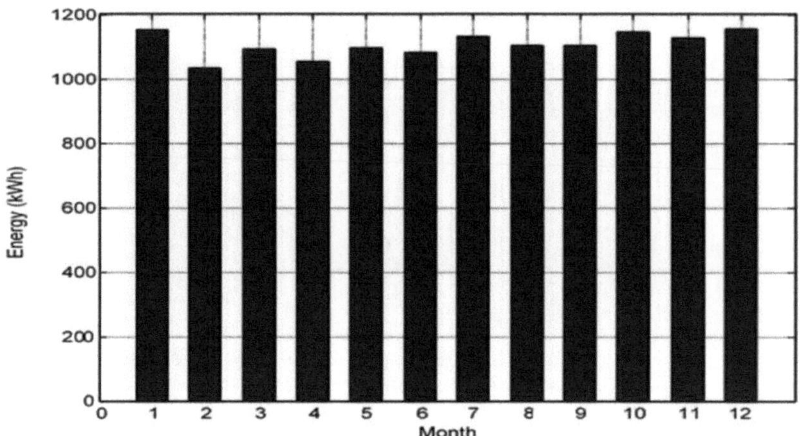

Figure-4-33 Monthly disposal energy in non-optimal mode

These figures show the amount of wasted energy that is greater in the colder months of the year, especially in winter because the micro turbine works longer hours to generate heat energy.

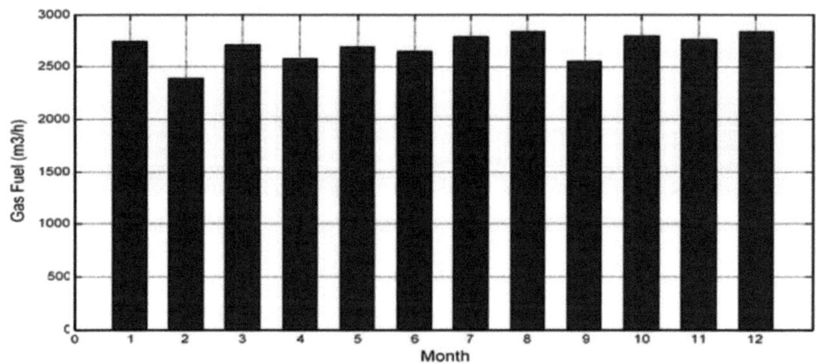

Figure-4-34 Optimum monthly consumption of micro turbine

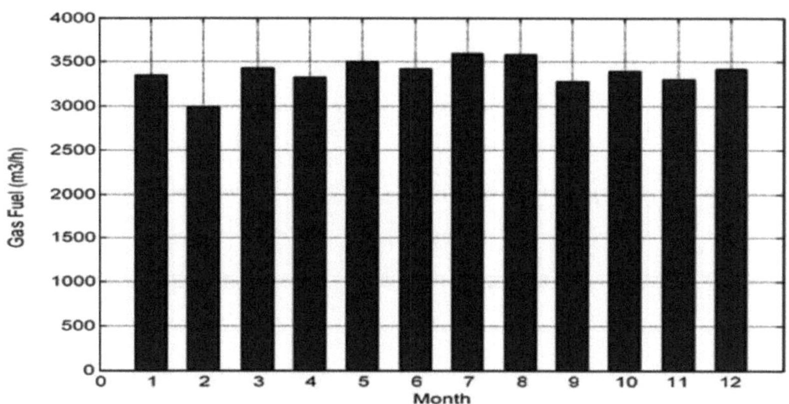

Figure-4-35 Monthly consumption of micro turbine in non-optimal state

These figures show the amount of fuel consumed by the micro turbine, which is comparable to that observed in the optimum state because of the greater use of the micro turbine.

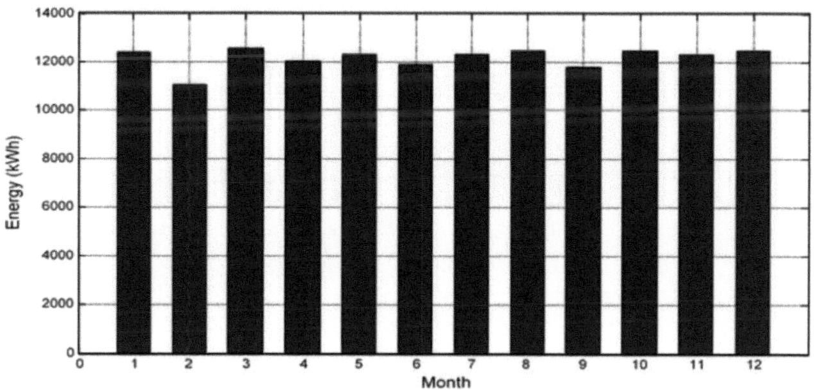

Figure-4-36 The difference between the total energy produced and the energy stored at optimum

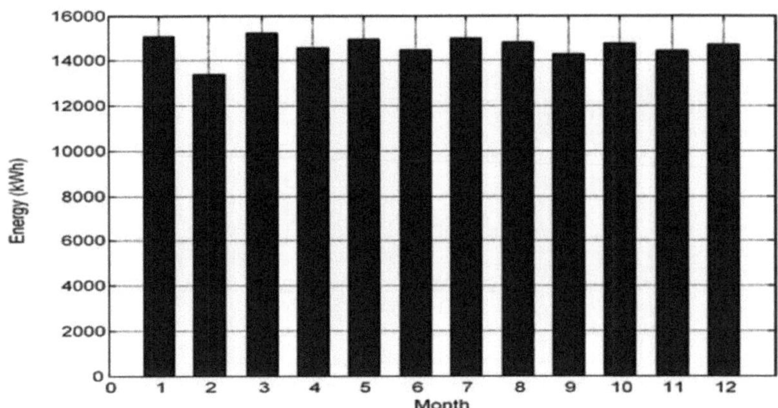

Figure-4-37 Total difference of energy produced and stored from energy consumed in non-optimal state

These figures show the difference between the total energy produced and the stored energy from the energy consumed. And that means we will never lack energy. Of course, this value is higher than the optimized state without optimization, which in turn indicates an additional cost to the non-optimized system.

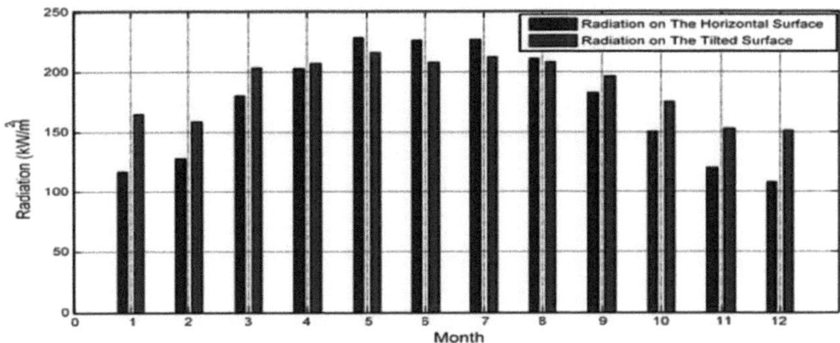

Figure-5-38 Optimized monthly radiation on the horizontal and tilted for one year in optimal condition

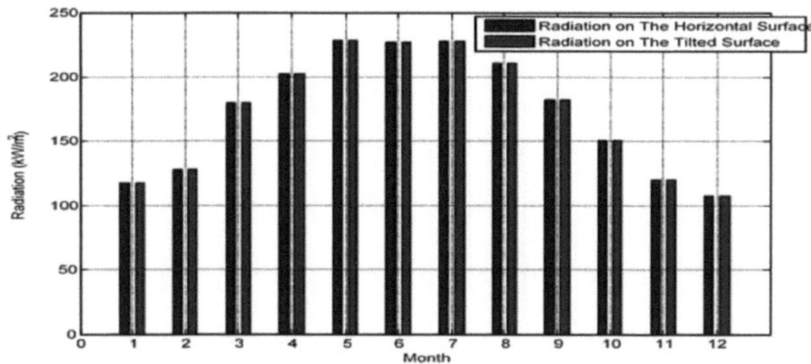

Figure-4-39 Optimized monthly radiation on the horizontal and tilted for one year in optimal condition

These figures show the solar radiation on the horizontal and sloping surfaces. When the slope was optimized, the maximum increase in energy received from sunlight was shifted to the colder months of the year and received more energy in the summer and early fall.

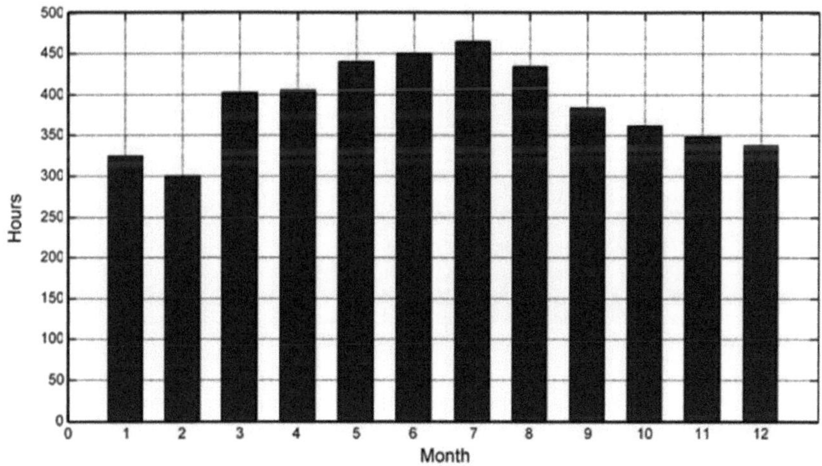

Figure-4-40 Sunny hours per month

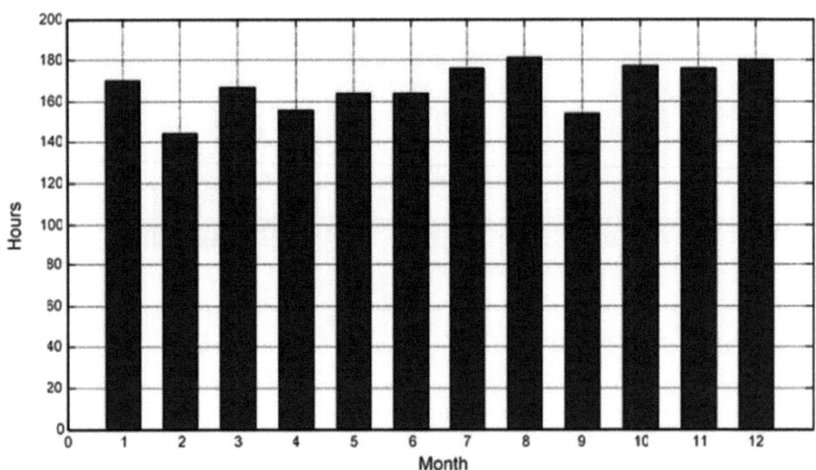

Figure-4-41 Micro turbine operating hours per month in optimal condition

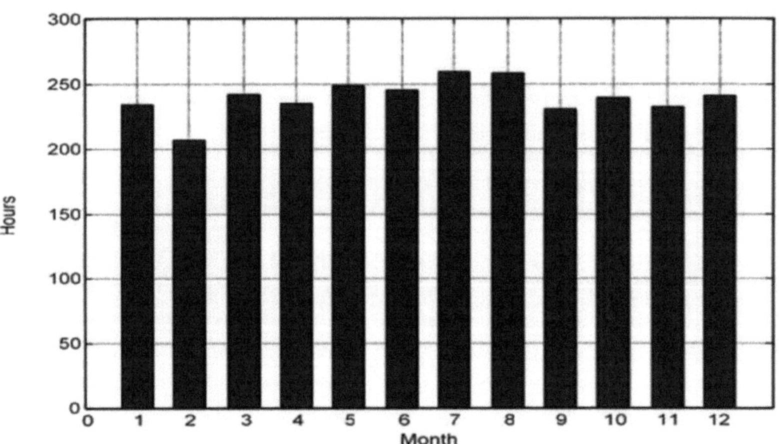

Figure-4-42 Micro turbine operating hours per month in non-optimal mode

This figure shows the micro turbine performance per month, which is lower in summer due to more sunlight and more in winter. Compared with the optimum mode, it is observed that in the optimum mode the number of micro turbines is lower.

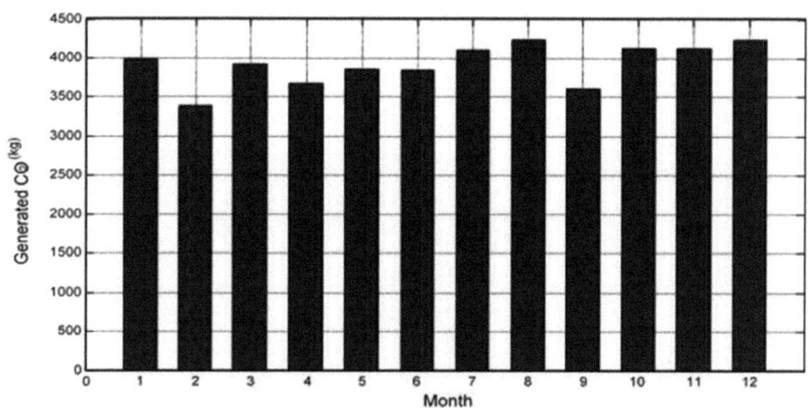

Figure-4-43 Micro turbine monthly production of carbon dioxide in optimum condition

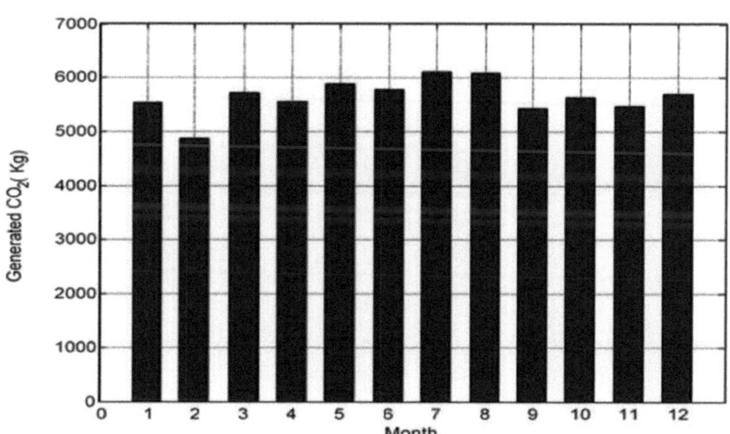

Figure-4-44 Micro turbine monthly carbon dioxide production in non-optimum
condition

These figures show the amount of carbon dioxide produced by the micro turbine, which in the optimum state is less contaminated than in the non-optimum state because the micro turbine performance is less optimal.

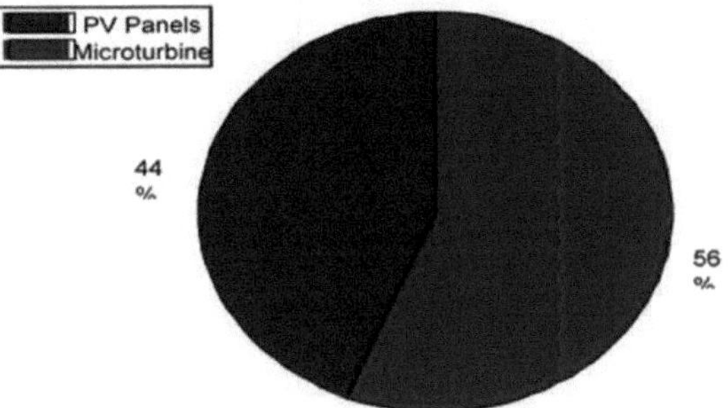

Figure-4-45 Solar Panel and Micro turbine Contribution in Optimal Yearly Energy Production

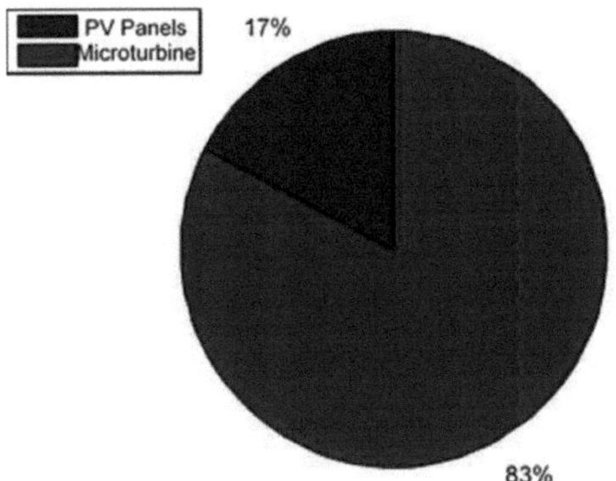

Figure-4-46 Solar Panel and Micro turbine Contribution to Annual Energy Outsourcing in non-optimum condition

This figure shows the contribution of solar panels and micro turbines. In the optimum case, the solar panel coefficient is about 44% and the micro turbine is 56%, while in the optimum case the values are 17% and 83%, respectively.

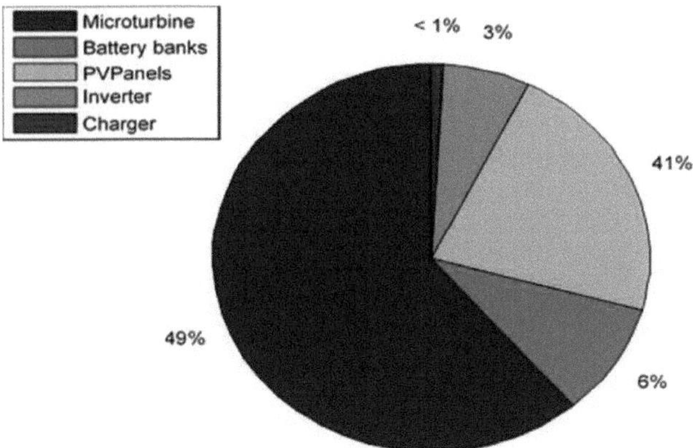

Figure-4-47 Cost of Hybrid System Components

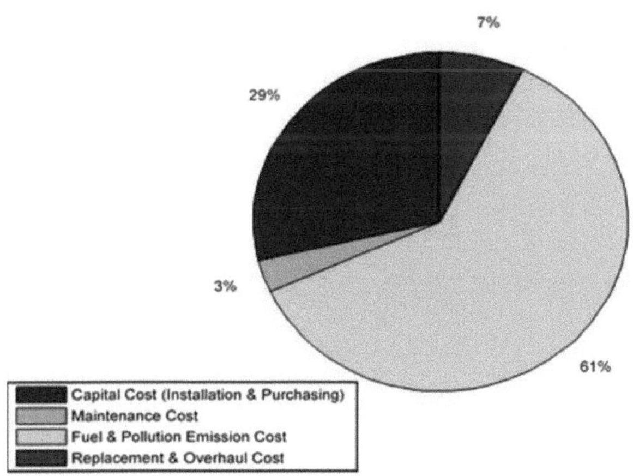

Figure-4-48 Cost of Micro turbine Components

This figure shows the cost of micro turbine components separately. These include installation and purchase costs (system initial cost), maintenance costs, fuel and pollution costs, repair and installation costs, and general overhaul.

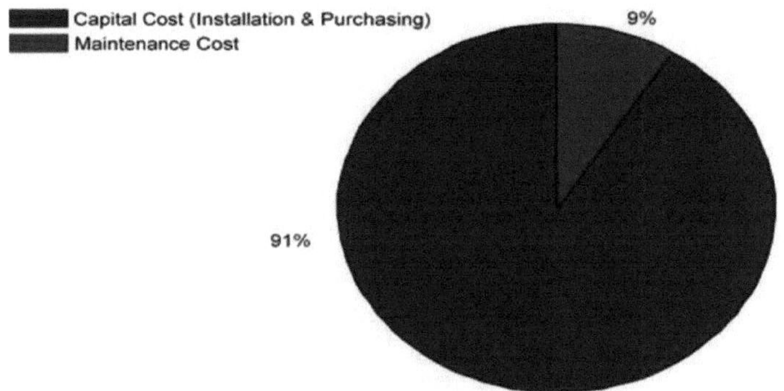

Figure-4-49 Cost of solar panel components

This figure shows the cost of solar panel components, including installation and purchase and maintenance costs. The solar panel has a lifespan of 25 years.

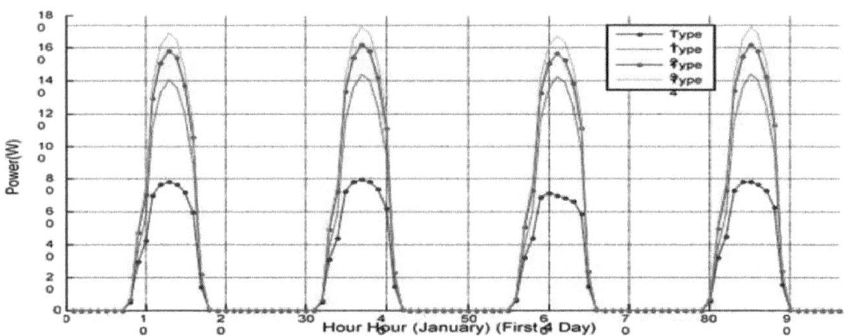

Figure-4-50 Power generation of 4 types of solar panels

In this book, we had 4 types of solar panels, all of which were examined and angles were optimized in all of them. It was then installed in the desired location and it was found that the fourth type panel had more capacity than the others.

Chapter V

Conclusions

Conclusions

The purpose of this book is to examine existing structures to optimize components of a hybrid system and thus minimize the cost of energy production. The feasibility of using a micro-turbine as a waiting source in hybrid power systems can be generalized. This conclusion is drawn from the fact that the feasibility of using a diesel generator as a waiting source in the hybrid power system, especially for other applications, has been substantiated by previous research and the possibility of using micro turbines as waiting sources instead of diesel generators for this purpose. The research is proven.

As previously shown, the optimized scenario is the micro-turbine / PV hybrid system. In fact, the research that analyzed the composition of the hybrid system used the micro-turbine as a backing source, one of the first published studies in this field. To the best of the authors' knowledge, no system of this kind has ever been installed in the world that publishes the results and can be used as an experimental benchmark for comparing the results of this research simulation.

In a time of optimized micro-turbine / PV scenario, the authors hoped that the results of the research would allow the Palestinian Authority to install this optimized PV micro-turbine scenario, the first project to be used for further analysis.

The purpose of this book is to model an optimal design of a PV micro-turbine / battery hybrid power system to provide power or power in remote Palestinian communities. The cost of disseminating global awareness is taken into account in this optimization analysis. The results of this analysis show that in this scenario, the hybrid system consists of a combination of PV plates, battery units and a micro-turbine which represents the most economical option as a source of expectation.

Here the micro turbine automates at measured power to supply the electric part of the load, while the micro turbine generating feature is used to store the heat load directly to make it economical. The COE, in this case, is 0.2598 / KWh. In a load curve with an average daily energy of 243 KWh and a peak power of 19 kW, it was observed that 92 panels (021.62kwp) and 20 battery units (25.92kwp) could supply this load with zero LLP. This micro-turbine, which has a rated power of 30KW, is required for commissioning as a backup source for the annual 2692h. Parts of these PV and micro turbine plates change from month to month.

The simulation results show that the generated energy is about 57% of the total energy produced by the micro turbine itself and the PV plate. These simulation results show that 4% of the total energy produced is discarded, while the total energy loss on an annual basis is about 14%. This means that the energy efficiency of this hybrid system is about 86%.

The cost of micro turbine capital is the highest among other component capital costs. About 49% of the initial costs are total, while about 41% of the total costs are for the PV system, in addition to the PV panels including the PV regulator and the battery bank. The rest of the cost goes to the friendship inverter. The energy divided by the micro turbine and PV system for each month is shown in Figure 5.2. The monthly charge and dissipation energies are also shown in the same figure. It is noteworthy that the energy produced by the micro-turbine is higher in the months with less PV energy. During these months, the waste energy is even higher. This is since the micro-turbine operates at its rated power while the heat consumed by the heat exchanger from the micro-turbine directly supplies the load heat energy. In the winter, the micro turbines work long hours, which is in line with the higher heat load directly supplied by the micro turbines. This explains more waste energy during the winter months.

The feasibility of using micro turbines as standby sources instead of diesel generators for this purpose has been demonstrated in this study. Diesel generators can be used effectively in hybrid renewable energy systems, but the small size of micro-turbines (<25KW) is not known in today's markets and the distribution of micro-turbines is limited.

The COE for residential uses and uses in Palestine is about 0.198 / KWh. The COE of the proposed hybrid system is higher than that of consumer companies. Implementation of these hybrid systems is highly acceptable if locations beyond the grid or emissions are considered. Also, prices for micro-turbines and PV system components are very acceptable and energy systems are expanding soon.

The economic analysis in this book uses the life cycle cost. This is when trying to compare different scenarios being analyzed to choose the least cost option. For comparison purposes, the cost of producing one unit of energy is calculated for each scenario. Economic analysis also needs to define the project life cycle time. It determines its life cycle with the maximum life span of the various components in each project. In the hybrid system considered in this book, the maximum life span for PV panels is 25 years. For each micro turbine, the manufacturer or manufacturer usually determines the time (in operating hours) before disassembly and the time (It also specifies the operating hours) before replacing the micro turbine. Battery life depends more on the number of battery discharge-charge cycles and the amount of charge-discharge depth.

References

[1] KR. Genwa, CP. Sagar. "Energy efficiency, solar energy conversion and storage in photogalvanic cell". Energy Convers Manage; 66:121–6, 2013.

[2] M. Alsayed, M. Cacciato, G. Scarcella, G. Scelba, "Multicriteria optimal sizing of photovoltaic-wind turbine grid connected systems". IEEE Trans Energy Convers, 28:370–9, 2013.

[3] MS. Ismail, M. Moghavvemi, TMI. Mahlia. "Techno-economic analysis of an optimized photovoltaic and diesel generator hybrid power system for remote houses in a tropical climate". Energy Convers Manage; 69:163–73, 2013.

[4] H. Meyar-Naimi, S. Vaez-Zadeh. "Sustainable development based energy policy making frameworks, a critical review". Energy Policy; 43:351–61, 2012.

[5] AN. Celik, T. Muneer. "Neural network based method for conversion of solar radiation data". Energy Convers Manage; 67:117–24, 2013.

[6] R. Saidur, G. BoroumandJazi, S. Mekhlif, M. Jameel. "Energy analysis of solar energy applications". Renew Sust Energy Rev; 16:350–6, 2012.

[7] R. Sen, SC. Bhattacharyya. "Off-grid electricity generation with renewable energy technologies in India: an application of HOMER". Renew energy; 62:388–98, 2014.

[8] FA. Mohamed, HN. Koivo. "Online management genetic algorithms of microgrid for residential application". Energy Convers Manage; 64:562–8, 2012.

[9] MS. Ismail, M. Moghavvemi, TMI. Mahlia. "Current utilization of microturbines as a part of a hybrid system in distributed generation technology". Renew Sust Energy Rev; 21:142–52, 2013.

[10] KH. Solangi, MR. Islam, R. Saidur, NA. Rahim, H. Fayaz. " A review on global solar energy policy". Renew Sust Energy Rev; 15:2149–63, 2011.

[11] JP. Torreglosa, P. García, LM. Fernández, F. Jurado. "Hierarchical energy management system for stand-alone hybrid system based on generation costs and cascade control". Energy Convers Manage; 77:514–26, 2014.

[12] SM. Shaahid, LM. Al-Hadhrami, MK. Rahman. "Review of economic assessment of hybrid photovoltaic–diesel–battery power systems for residential loads for different provinces of Saudi Arabia". Renew Sust Energy Rev; 31: 174-81, 2014.

[13] MS. Ismail, M. Moghavvemi, TMI. Mahlia. "Analysis and evaluation of various aspects of solar radiation in the Palestinian Territories". Energy Convers Manage; 73: 57–68, 2013.

[14] Publications – Energy – Household Energy Survey: Main Results (January). Palestinian Central Bureau of, Statistics (PCBS); 2011.

[15] MS. Ismail, M. Moghavvemi, TMI. Mahlia. "Characterization of PV panel and global optimization of its model parameters using genetic algorithm". Energy Convers Manage; 73: 10–25, 2013.

[16] O, Erdinc, M. Uzunoglu. "Optimum design of hybrid renewable energy systems: overview of different approaches". Renew Sust Energy Rev; 16:1412–25, 2012.

[17] E. Koutroulis, D. Kolokotsa, A. Potirakis, K. Kalaitzakis. "Methodology for optimal sizing of stand-alone photovoltaic/wind-generator systems using genetic algorithms". Solar Energy; 80:1072–88, 2006.

[18] Y. Hongxing, Z. Wei, L. Chengzhi. "Optimal design and techno-economic analysis of a hybrid solar–wind power generation system". Appl Energy; 86:163–9, 2009.

[19] R. Dufo-Lopez, JL. Bernal-Agustin. "Multi-objective design of PV–wind–diesel–hydrogen–battery systems". Renew Energy; 33:2559–72, 2008.

[20] RK. Rajkumar, VK. Ramachandaramurthy, BL. Yong, DB. Chia. "Techno-economical optimization of hybrid pv/wind/battery system using Neuro-Fuzzy". Energy; 36:5148–53, 2011.

[21] W. Caisheng, CM. Colson, MH. Nehrir, L. Jian. "Power management of a standalone hybrid wind-microturbine distributed generation system". In: Powerelectronics and machines in wind applications, PEMWA, IEEE, p. 1–7, 2009.

[22] MS, Ismail, M, Moghavvemi, TMI, Mahlia. "Design of an optimized photovoltaic and micro turbine hybrid power system for a remote small community: case study of Palestine". Energy Convers Manage; 75:271–81, 2013.

[23] JA. Duffie, WA. Beckman. "Solar engineering of thermal processes". 3rd ed. San Francisco: Wiley; 2006.

[24] M. Bortolini, M. Gamberi, A. Graziani, C. Mora, A. Regattieri. "Multi-parameter analysis for the technical and economic assessment of photovoltaic systems in the main European Union countries". Energy Convers Manage; 74:117–28, 2013.

[25] DG. Lorente, S. Pedrazzi, G. Zini, A. Dalla Rosa, P. Tartarini. "Mismatch losses in PV power plants". Solar Energy; 100:42–9, 2014.

[26] H. Kord, A. Rohani. "An integrated hybrid power supply for off-grid applications fed by wind/photovoltaic/fuel cell energy systems". In: 24 h Internationa power system conference; p. 1–11,2009.

[27] Combined Heat and Power Partnership. Catalog of CHP technologies. U.S. Environmental Protection Agency; p. 139, 2008.

[28] S. Vaez-Zadeh, AH. Isfahani. "Multiobjective design optimization of air-core linear permanent-magnet synchronous motors for improved thrust and low magnet consumption". Magnetics, IEEE Transactions on, vol. 42; p. 446–52, 2006.

[29] Whole Solar Website; 2013.

[30] M. Lalwani, DP. Kothari, M. Singh. "Investigation of solar photovoltaic simulation softwares". Int J Appl Eng Res, Dindigul;1:585–601, 2010.

[31] T. Khatib, A. Mohamed, K. Sopian. " A software tool for optimal sizing of PV systems in Malaysia". Model Simul Eng: 1–11, 2012.

[32] AMI. Yasin. " Optimal operation strategy and economyic analysis of rural electrification of Atouf village by electric network, diesel generator and photovoltaic system (Master thesis) ". Nablus: Electrical Engineering Department, An-Najah National University; 2008

[33] Kashif, Ishaque, Zainal, Salam, "an improved modeling method to determine the model parameters of photovoltaic (PV) modules using differential evolution (DE) ", Solar Energy; 85:2349–2359, 2011

Printed by Books on Demand GmbH, Norderstedt / Germany